Understanding Lasers

Understanding Lasers

Christian West

MURPHY & MOORE
www.murphy-moorepublishing.com

Published by Murphy & Moore Publishing
1 Rockefeller Plaza,
New York City, NY 10020, USA
www.murphy-moorepublishing.com

Understanding Lasers
Christian West

International Standard Book Number: 978-1-63987-554-2 (Hardback)

Cataloging-in-Publication Data

Understanding lasers / Christian West.
p. cm.
Includes bibliographical references and index.
ISBN 978-1-63987-554-2
1. Lasers. 2. Lasers in engineering. 3. Light sources.
4. Nonlinear optics. I. West, Christian.
TA1675 .U53 2022
621.366--dc23

Contents

Preface

Laser stands for light amplification by stimulated emission of radiation. A laser is any device that emits a narrow beam of light at a particular wavelength by stimulating atoms and molecules using electromagnetic radiation. Various components of a laser beam device are gain medium, high reflector, laser pumping energy, laser beam and an output coupler. A laser beam profiler is used to measure the depth, divergence and intensity profile of laser beams. Lasers are highly coherent in nature. This allows them to stay focused in a tight spot, making lasers an excellent choice for use in various domains such as printing, welding, cutting, photolithography, weapons, fiber optics, DNA sequencing, etc. Lasers are of various types such as solid-state lasers, gas lasers, dye lasers, semiconductor lasers, etc. Use of laser in medical treatments such as laser hair removal, laser mammography is known as laser medicine. Lasers are used in either continuous wave operation or pulsed operation. This book elucidates the concepts and innovative models around prospective developments with respect to lasers. It is compiled in such a manner, that it will provide in-depth knowledge about the theories related to lasers. Coherent flow of topics, student-friendly language and extensive use of examples make this book an invaluable source of knowledge.

A foreword of all chapters of the book is provided below:

Chapter 1 - Laser is a device which emits light through optical amplification which is based on stimulated emission of electromagnetic radiation. A few major types of lasers are solid state laser, gas laser, dye laser, semiconductor laser and free electron laser. This chapter will provide a concise introduction to the significant aspects of these types of lasers; Chapter 2 - The working of laser depends on various theories such as interaction of light with matter, absorption and emission of light, Plank's quantum theory of light, Einstein coefficient relation, laser rate equation, population inversion and optical pumping. The diverse aspects and applications of these theories in relation to lasers have been thoroughly discussed in this chapter; Chapter 3 - Machining involves cutting a material into a desired final shape and size. Laser machining is a thermal machining process wherein a laser beam is used for the production of heat which removes material from a workpiece. The topics elaborated in this chapter will help in gaining a better perspective about laser machining as well as laser beam welding and laser shock peening; Chapter 4 - The technology which makes use of lasers in order to vaporize materials for the purpose of producing a cut edge is known as laser cutting technology. A few different methods of laser cutting are vaporization cutting, stealth dicing and thermal stress cracking. The topics elaborated in this chapter will help in gaining a better perspective about laser cutting as well as these types of laser cutting methods; Chapter 5 - Lasers have a wide variety of uses in various industries and fields. A few of these are optical disk drive, laser printing, 3D scanner, laser surgery, fiber-optic communication and laser pointer. This chapter closely examines the applications of lasers in these fields to provide an extensive understanding of the subject.

At the end, I would like to thank all the people associated with this book devoting their precious time and providing their valuable contributions to this book. I would also like to express my gratitude to my fellow colleagues who encouraged me throughout the process.

Christian West

1

Laser and its Types

Laser is a device which emits light through optical amplification which is based on stimulated emission of electromagnetic radiation. A few major types of lasers are solid state laser, gas laser, dye laser, semiconductor laser and free electron laser. This chapter will provide a concise introduction to the significant aspects of these types of lasers.

Lasers are amazing light beams powerful enough to zoom miles into the sky or cut through lumps of metal. Although they seem pretty recent inventions, they've actually been with us over half a century: the theory was figured out in 1958; the first practical laser was built in 1960. At that time, lasers were thrilling examples of cutting-edge science: secret agent 007, James Bond, was almost chopped in half by a laser beam in the 1964 film Gold finger. But apart from Bond villains, no-one else had any idea what to do with lasers; famously, they were described as "a solution looking for a problem." Today, we all have lasers in our homes (in CD and DVD players), in our offices (in laser printers), and in the stores where we shop (in barcode scanners). Our clothes are cut with lasers, we fix our eyesight with them, and we send and receive emails over the Internet with signals that lasers fire down fiber-optic cables. Whether we realize it or not, all of us use lasers all day long, but how many of us really understand what they are or how they work?

The basic idea of a laser is simple. It's a tube that concentrates light over and over again until it emerges in a really powerful beam. But how does this happen, exactly? What's going on inside a laser?

What is a Laser?

Lasers are more than just powerful flashlights. The difference between ordinary light and laser light is like the difference between ripples in your bathtub and huge waves on the sea. You've probably noticed that if you move your hands back and forth in the bathtub you can make quite strong waves. If you keep moving your hands in step with the waves you make, the waves get bigger and bigger. Imagine doing this a few million times in the open ocean. Before long, you'd have mountainous waves towering over your head! A laser does something similar with light waves. It starts off with weak light and keeps adding more and more energy so the light waves become ever more concentrated.

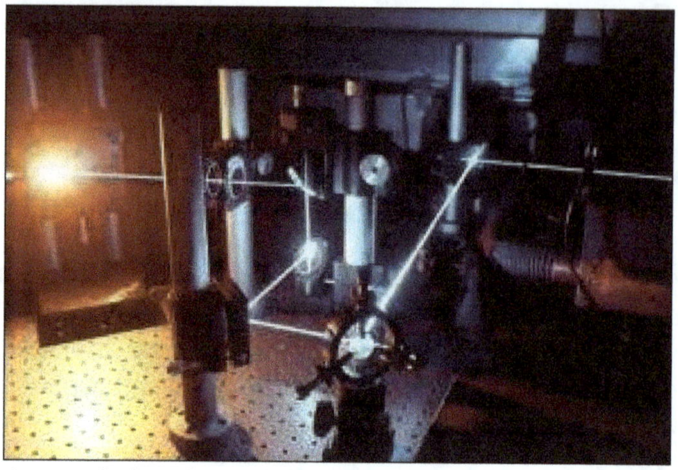

It's much easier to make laser beams follow precise paths than ordinary light beams, as in this experiment to develop better solar cells.

If you've even seen a laser in a science lab, you'll have noticed two very important differences straightaway:

- Where a flashlight produces "white" light (a mixture of all different colors, made by light waves of all different frequencies), a laser makes what's called monochromatic light (of a single, very precise frequency and color—often bright red or green or an invisible "color" such as infrared or ultraviolet).
- Where a flashlight beam spreads out through a lens into a short and fairly fuzzy cone, a laser shoots a much tighter, narrower beam over a much longer distance (we say it's highly collimated).

There's a third important difference you won't have noticed:

- Where the light waves in a flashlight beam are all jumbled up (with the crests of some beams mixed with the troughs of others), the waves in laser light are exactly in step: the crest of every wave is lined up with the crest of every other wave. We say laser light is coherent. Think of a flashlight beam as a crowd of commuters, pushing and shoving, jostling their way down the platform of a railroad station; by comparison, a laser beam is like a parade of soldiers all marching precisely in step.

These three things make lasers precise, powerful, and amazingly useful beams of energy.

How Lasers Work?

A laser is effectively a machine that makes billions of atoms pump out trillions of photons (light particles) all at once so they line up to form a really concentrated light beam.

What do you Need to Make a Laser?

We need two basic parts:

- A load of atoms (a solid, liquid, or gas) with electrons in them that we can stimulate. This is known as the medium or, sometimes, the amplifying or "gain" medium (because gain is another word for amplification).
- Something to stimulate the atoms with, such as a flash tube (like the xenon flash lamp in a camera) or another laser.

A typical red laser would contain a long crystal made of ruby (the medium, shown in the artwork below as a red bar) with a flash tube (yellow zig-zag lines) wrapped around it. The flash tube looks a bit like a fluorescent strip light, only it's coiled around the ruby crystal and it flashes every so often like a camera's flash lamp.

How do the Flash Tube and the Crystal Make a Laser Beam?

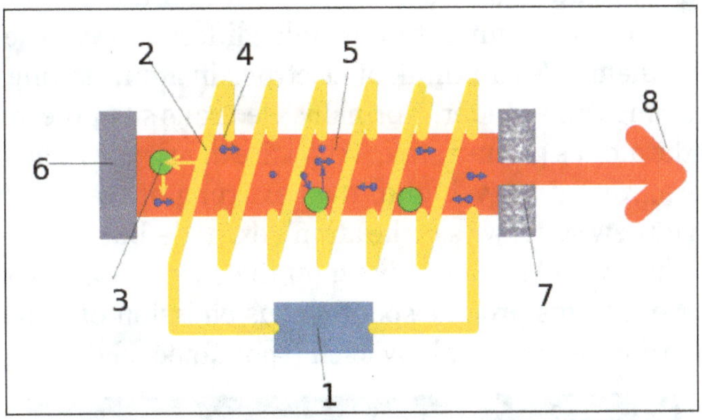

A high-voltage electric supply makes the tube flash on and off. Every time the tube flashes, it "pumps" energy into the ruby crystal. The flashes it makes inject energy into the crystal in the form of photons. Atoms in the ruby crystal (large green blobs) soak up this energy in a process called absorption. Atoms absorb energy when their electrons jump to a higher energy level. After a few milliseconds, the electrons return to their original energy level (ground state) by giving off a photon of light (small blue blobs). This is called spontaneous emission.

The photons that atoms give off zoom up and down inside the ruby crystal, traveling at the speed of light. Every so often, one of these photons stimulates an already excited atom. When this happens, the excited atom gives off a photon and we get our original photon back as well. This is called stimulated emission. Now one photon of light has produced two, so the light has been amplified (increased in strength). In other words, "light amplification"(an increase in the amount of light) has been caused by "stimulated emission of radiation" (hence the name "laser", because that's exactly how a laser works).

A mirror at one end of the laser tube keeps the photons bouncing back and forth inside the crystal. A partial mirror at the other end of the tube bounces some photons back into the crystal but lets some escape. The escaping photons form a very concentrated beam of powerful laser light.

How do Lasers Make Light?

"Laser" stands for Light Amplification by Stimulated Emission of Radiation. That's a complex and confusing mouthful but, if you pull it apart slowly, it's actually a very clear explanation of how lasers make their super-powerful beams of light.

Spontaneous Emission

Let's start with the "R" of laser: radiation. The radiation lasers make has nothing to do with dangerous radioactivity, the stuff that makes Geiger counters click, which atoms spew out when they smash together or fall apart. Lasers make electromagnetic radiation, just like ordinary light, radio waves, X rays, and infrared. Although it's still produced by atoms, they make ("emit") it in a totally different way, when electrons jump up and down inside them. We can think of electrons in atoms sitting on energy levels, which are a bit like rungs on a ladder. Normally, electrons sit at the lowest possible level, which is called the atom's ground state. If you fire in just the right amount of energy, you can shift an electron up a level, onto the next rung of the "ladder." That's called absorption and, in its new state, we say the atom's excited—but it's also unstable. It very quickly returns to the ground state by giving off the energy it absorbed as a photon (a particle of light). We call this process spontaneous emission of radiation: the atom is giving off light (emitting radiation) all by itself (spontaneously).

From candles to light bulbs and fireflies to flashlights, all conventional forms of light work through the process of spontaneous emission. In a candle, combustion (the chemical reaction between oxygen and fuel, in this case, wax) excites atoms and makes them unstable. They give off light when they return to their original (ground) states. Every photon produced by spontaneous emission inside this candle flame is different from every other photon, which is why there's a mixture of different wavelengths (and colors), making "white" light. The photons emerge in random directions, with waves that are out of step with one another ("out of phase"), which is why candlelight is much weaker than laser light.

Stimulated Emission

Normally, a typical bunch of atoms would have more electrons in their ground states than their excited states, which is one reason why atoms don't spontaneously give off light. But what if we excited those atoms—pumped them full of energy—so their electrons were in excited states. In that case, the "population" of excited electrons would be bigger than the "population" in their ground states, so there would be plenty of electrons ready and willing to make photons of light. We call this situation a population inversion, because the usual state of affairs in the atoms is swapped around (inverted). Now suppose also that we could maintain our atoms in this state for a little while so they didn't automatically jump back down to their ground state (a temporarily excited condition known as a meta-stable state). Then we'd find something really interesting. If we fired a photon with just the right energy through our bunch of atoms, we'd cause one of the excited electrons to jump back down to its ground state, giving off both the photon we fired in and the photon produced by the electron's change of state. Because we're stimulating atoms to get radiation out of them, this process is called stimulated emission. We get two photons out after putting one photon in, effectively doubling our light and amplifying it (increasing it). These two photons can stimulate other atoms to give off more photons, so, pretty soon, we get a cascade of photons—a chain reaction—throwing out a brilliant beam of pure, coherent laser light. What we've done here is amplify light using stimulated emission of radiation—and that's how a laser gets its name.

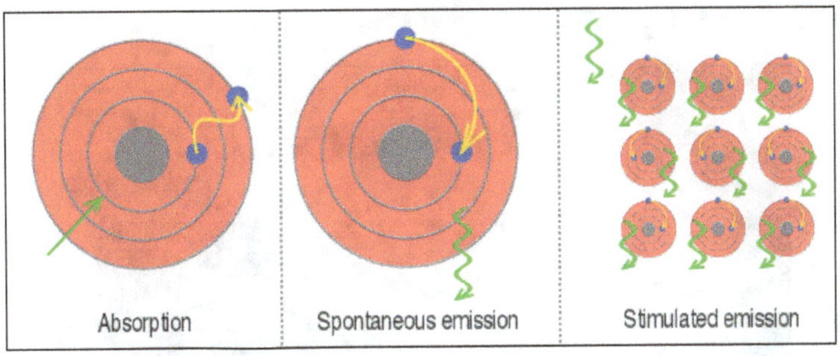

How lasers work in theory: Left: Absorption: Fire energy (green) into an atom and you can shift an electron (blue) from its ground state to an excited state, which usually means pushing it further from the nucleus (gray). Middle: Spontaneous emission: An excited electron will naturally jump back to its ground state, giving out a quantum (packet of energy) as a photon (green wiggle). Right: Stimulated emission: Fire a photon near a bunch of excited atoms and you can trigger a cascade of identical photons. One photon of light triggers many, so what we've got here is light amplification (making more light) by stimulated emission of (electromagnetic) radiation—laser.

What Makes Laser Light so Different?

If that's how lasers make light, why do they make a single color and a coherent beam? It boils down to the idea that energy can only exist in fixed packets, each of which is called a quantum. It's a bit like money. You can only have money in multiples of the most basic unit of your currency, which might be a cent, penny, rupee, or whatever. You can't have a tenth of a cent or a twentieth of a rupee, but you can have 10 cents or 20 rupees. The same is true of energy, and it's particularly noticeable inside atoms.

Like the rungs on a ladder, the energy levels in atoms are in fixed places, with gaps in between them. You can't put your foot anywhere on a ladder, only on the rungs; and in exactly the same way, you can only move electrons in atoms between the fixed energy levels. To make an electron jump from a lower to a higher level, you have to feed in a precise amount (quantum) of energy, equal to the difference between the two energy levels. When electrons flip back down from their excited to their ground state, they give out the same, precise amount of energy, which takes the form of a photon of light of a particular color. Stimulated emission in lasers makes electrons produce a cascade of identical photons—identical in energy, frequency, wavelength—and that's why laser light is monochromatic. The photons produced are equivalent to waves of light whose crests and troughs line up (in other words, they are "in phase")—and that's what makes laser light coherent.

What are Lasers used for?

When Theodore Maiman developed the first practical laser, few people realized how important these machines would eventually become. Over half a century later, applications like this—precision tools, digital communication, and defense—remain among the most important uses of lasers.

Every time it prints a document, the laser printer on your desk is busily stimulating zillions of atoms. The laser inside it is used to draw a very precise image of the page you want to print onto a large drum, which picks up powered ink (toner), and transfers it onto paper.

Tools

Cutting tools based on CO_2 lasers are widely used in industry: they're precise, easy-to-automate, and, unlike knives, never need sharpening. Where pieces of cloth were once cut by hand to make things like denim jeans, now fabrics are chopped by robot-guided lasers. They're faster and more accurate than humans and can cut multiple thicknesses of fabric at once, which improves efficiency and productivity. The same precision is equally important in medicine: doctors routinely use lasers on their patients' bodies. For everything from blasting cancer tumors and cauterizing blood vessels to correcting problems with people's vision (laser-eye surgery, fixing detached retinas, and cataract treatments all involve lasers).

Communications

Lasers form the bedrock of all kinds of 21st-century digital technology. Every time you swipe your shopping through a grocery store barcode scanner, you're using a laser to convert a printed barcode into a number that the checkout computer can understand. When you watch a DVD or listen to a CD, a semiconductor laser beam bounces off the spinning disc to convert its printed pattern of data into numbers; a computer chip converts these numbers into movies, music, and sound. Along with fiber-optic cables, lasers are widely used in a technology called photonics—using photons of light to communicate, for example, to send vast streams of data back and forth over the Internet. Facebook is currently experimenting with using lasers (instead of radio waves) to make better connections to space satellites, which could lead to higher data rates and much-improved Internet access in developing countries.

Are laser weapons the future? This is the US Navy's Laser Weapon System (LaWS), which was tested onboard the USS Ponce in 2014. There are no expensive bullets or missiles with a laser gun like this, just an endless supply of fiercely directed energy.

Defense

The military has long been one of the biggest users of this technology, mainly in laser-guided weapons and missiles. Despite its popularization in movies and on TV, the sci-fi idea of laser weapons that can cut, kill, or blind an enemy remained fanciful until the mid-1980s. In 1981, The New York Times went so far as to quote one "military laser expert" saying: "It's just silly. It takes more energy to kill a single man with a laser than to destroy a missile." Two years later, long-range laser weapons officially became the bedrock of US President Ronald Reagan's controversial Strategic Defense Initiative (SDI), better known as the "Star Wars program". The original idea was to use space-based, X ray lasers (among other technologies) to destroy incoming enemy missiles before they had time to do damage, though the plan gradually fizzled out following the collapse of the Soviet Union and the end of the Cold War.

Even so, defense scientists have continued to transform laser-based missiles from science fiction into reality. The US Navy first began testing LaWS (Laser Weapon System) onboard the USS Ponce ship in the Persian Gulf in 2014. Using solid-state lasers pumped by LEDs, it was designed to damage or destroy enemy equipment more cheaply and precisely than conventional missiles. The tests proved successful, and the Navy announced contracts to build more LaWS systems in 2018. Meanwhile, the development of space lasers continues, though none have so far been deployed.

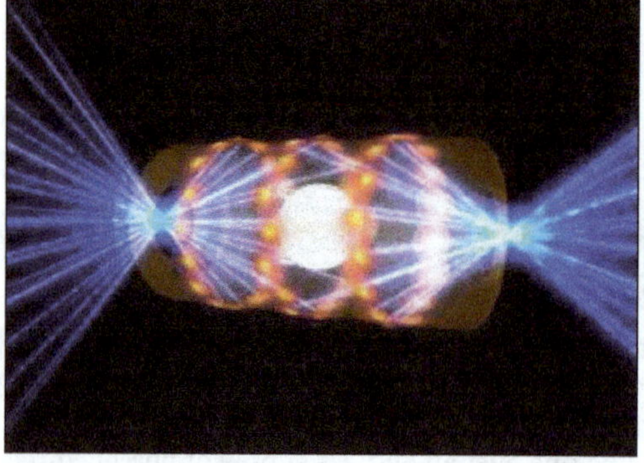

Scientists at Lawrence Livermore National Laboratory in California developed the world's most powerful laser, National Ignition Facility (NIF), for nuclear research. Housed in a 10-story building occupying an area as big as three football fields, it uses 192 separate laser beams to deliver up to 500 trillion watts of power (100 times more energy than any other laser), generating temperatures of up to 100 million degrees. NIF cost a total of $3.5 billion and is expected to power cutting-edge nuclear research for the next 30 years. Left: One of the twin laser bays at the National Ignition Facility. Right: How it works: Beams from the laser are concentrated on a small pellet of fuel in a chamber to produce intense temperatures (like those deep inside stars). The idea is to produce nuclear fusion (make atoms join together) and release a massive amount of energy.

Discrete Thermodynamics in Laser

The ideal laser system allows to pump as much as possible amount of excited atoms from their ground state(subsystem A) to the upper level (subsystem A*), then holds them over there as long as necessary until the excited atoms get back to the ground state due to the light emission.

Suppose that laser contains 1 mole of the excitable atoms. At zero pumping force the laser stands in internal thermodynamic equilibrium (TdE), i.e. thermodynamic equilibrium between the subsystems A and A*. Their equilibrium populations are η (<1) moles of the A atoms and $(1-\eta)$ moles of the A* atoms; the ratio between them is defined by the Boltzmann distribution:

$$(1-\eta)/\eta = \exp(-h\nu/kT).$$

In open laser system the upper level population increases due to the energy pumping and absorption:

$$A + \varepsilon_{in} \rightarrow A^*,$$

While stimulated or spontaneous discharge of the accumulated energy via light emission:

$$A^* \rightarrow A + h\nu,$$

returns the excited laser population back to the ground.

Neglecting the heating of laser body x due to energy dissipation that may be prevented by cooling, lasers work at permanent pressure and temperature. Characteristic function in this case is Gibbs' free energy. In the laser system its change for the A→A* transition is: $\Delta G = \Delta G^0 + RT\ln[(1-\eta)/\eta]$.

At TdE, in absence of the pumping force ΔG=0 combination of $(1-\eta)/\eta = \exp(-h\nu/kT)$. and $\Delta G = \Delta G^0 + RT\ln\left[(1-\eta)/\eta\right]$,

leads to,

$$(1-\eta)/\eta = \exp\left(-\Delta G^0 / RT\right).$$

Multiplying both tiers of the exponent power in $(1-\eta)/\eta = \exp(-h\nu/kT)$. by the Avogadro number NA one can get expression for the standard change of Gibbs' free energy value, reduced by RT,

$$\Delta g0 = h\nu NA / RT,$$

or, after gathering the universal constants into λ_v=hN_A/R,

$$\Delta g^0 = \lambda_v \nu / T,$$

λ_v=47.99 if the frequency is expressed in terahertz. Now the amount of the ground level atoms at TdE is:

$$\eta = 1/\left[1+\exp\left(-\lambda_v \nu / T\right)\right].$$

For the most important part of the light spectrum, from far infrared to short wave UV, the Δg^o values vary from −63.040 for to −246.720 dimensionless units at T=300K. It is well known that at TdE the η values are very close to unity, i.e. the huge majority of the excitable laser dwellers are sitting on the ground as it takes place in real devices. On the other hand, that makes all frequencies within the above IR-UV range thermodynamically undistinguishable.

Thermodynamic Equation of State of the Laser System

If x moles of atoms were pumped off the ground level up, the laser system state proximity to TdE may be characterized by the system state dimensionless coordinate:

$$\Delta_e = (\eta - x)/\eta.$$

Obviously at TdE x=0 and Δ_e =1. We define the shift from TdE as δ_e = 1− Δ_e,

$$\delta_e = x/\eta.$$

At TdE Δ_e = 1, and δ_e=0. When the whole population is pumped up, Δ_e and δ_e exchange their values. The laser system is driven by the pumping force out of TdE until a new stationary state is achieved. From basic definition of equilibrium as a balance of acting against the system internal and external thermodynamic forces (TdF). The balance is given by Onsager equation:

$$A_i + a_{ij}P_j = 0,$$

The first term is well known thermodynamic affinity, which in discrete version is $A_i=-\Delta g/\Delta_e$. The second term contains reduced Onsager coefficient $a_{ij}=a_{ij}/a_{ii}$ and external, pumping thermodynamic force that equals to $P_j= -E_p/(RT)$, where Ep is the energy absorbed by 1 mole of the excitable atoms. In our case T=300K and RT=2.44 kJ/mol. Taking into account that relation, after through multiplication by Δe and substitution $a_{ij} = a_{ij}/RT$ we turn equation $A_i + a_{ij}P_j = 0$, to:

$$\Delta g - a_{ij}(1-\delta_e)E_p = 0.$$

To obtain the laser system equation of state from this precursor, we have to eliminate external factors leaving only variables and internal parameters. Let us express the pumping energy via the laser system reaction to pumping as a power series of the system shift from TdE, which in a simple case may look like:

$$E_p = \Sigma_{0 \le k \le q} w_k \delta_e^{\,k}$$

q is a loosely defined system complexity parameter. In case of laser system it may be identified roughly with the number of possible transitions between the levels, i.e. two in case of 2-level equilibrium (one up and one down). Let's move wo out of the sum in $E_p = \Sigma_{0 \le k \le q} w_k \delta_e^{\,k}$; new weights $wk_0 = w_k/w_0$ are still unknown and we have to set them to unities. It's very easy to see that after placing expression $E_p = \Sigma_{0 \le k \le q} w_k \delta_e^{\,k}$ for Ep into $\Delta g - a_{ij}(1-\delta_e)E_p = 0.$ and after multiplication $(1-\delta_e)$ by $(1+\delta_e^{\,1} + \delta_e^{\,2} + ... + \delta_e^{\,q})$, all powers of δ_e but q annihilate, and $\Delta g - a_{ij}(1-\delta_e)E_p = 0$ turn to the final equation of state for the open laser system as a logistic map:

$$\Delta g - \alpha(1-\delta_e^{\,q}) = 0,$$

where new coefficient is α=woaij. This laconic logistic map may be extended using $(1-\eta)/\eta = \exp(-\Delta G^0 / RT)$. and taking into account the population numbers, $\eta(1-\delta_e)$ on the ground and $[1-\eta(1-\delta_e)]$ on the upper level:

$$-\ln[(1-\eta)/\eta] + \ln\{[1-\eta(1-\delta_e)]/\eta(1-\delta_e)\} - \alpha(1-\delta_e^{\,q}) = 0.$$

Map above equation contains one variable δe and two parameters of the new theory, the above introduced q and α; the latter is known as the growth parameter in the theory of populations. In our approach it defines the rate of the laser system shift from TdE by the pumping force. Before proceeding with solutions, we'd like to mention one interesting feature of this map. It was already mentioned that η≈1; such approximation in above equation in combination with $(1-\eta)/\eta = \exp(-h\nu / kT)$ gives us a variation of the basic map:

$$_{v}\nu / T + \ln[\delta_e / (1-\delta_e)] - (1-\delta_e) = 0.$$

Let's neglect the A*–population at TdE (1–η≈0!); then, if x moles of A moved to A* the population ratio is π=x/(η–x); substituting expression for x from $\delta_e = x/\eta$. we get π=δη/(1–δη), or π=δ/(1–δ). That changes expression shown above to more explicit:

$$\lambda_v \nu / T + \ln\left[\delta_e / (1-\delta_e)\right] - \alpha\left(1-\delta_e^{\,q}\right) = 0.$$

All logistic maps are solved usually via iterations.

Solutions to the Laser Logistic MAP

We simulated open equilibria of the 2-level lasers, simulation results were graphically interpreted in Figure using dynamic inverse bifurcation diagrams, that is shift δ_e vs. pumping force P_j. Both values are dimensionless.

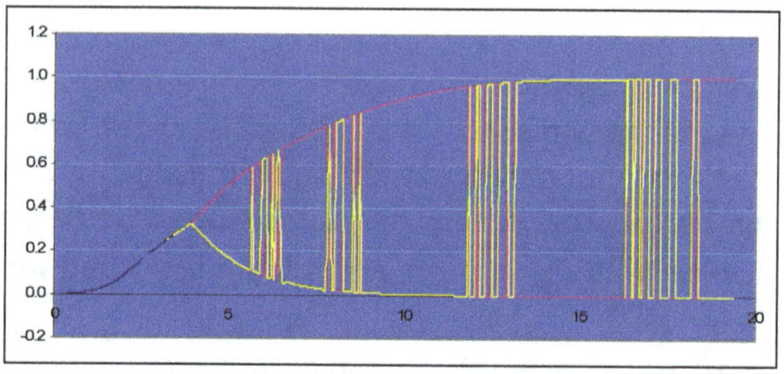

DIBD for the 2-level laser system at q=2.

As the pumping force and amount of activated atoms increase, the system equilibrium shifts from TdE following the thermodynamic branch. The branch looses stability at certain value of the force, splits by two bifurcation branches, and we encounter a bi-stability. Vertical lines mean instability of bifurcation lines; in real laser, downward lines on figure correspond to spontaneous transitions, i.e. spontaneous emission.

Figure shows short wave UV dynamic diagram, stretched horizontally. All the transitions frequency spectra, constituted by the immediately following each other up and down transitions are line spectra within the studied light frequency range. The ascending transitions are driven by external pumping power, bringing a part of the laser population to the upper level. Being pumped up, the excited atoms cannot accumulate more energy from the pumping flow: a stress, caused by the pumping "pressure", forces the system to discharge itself by spontaneous or assisted irradiation of light along the descending line. This result represents an independent proof of the Einstein's prohibition on the 2-level laser, based exclusively on discrete thermodynamics and extended to open laser systems.

It was found impossible in this research to get rid of the lines on the pumping force-system shift diagrams by changing the simulation parameters - this linearity seems to be imminent to the model of laser, offered by this theory as well as spontaneous emission in lasers at all. At the accepted simulation parameters, the first linearity starts at $\delta_e \approx$

0.6. If $\eta \approx 1$ that leads to $\pi \approx 0.45$, the value that is more than two times lower than population inversion threshold. Results prompt us to state that it is definitely very costly and ineffective to create and maintain population inversion in 2–level laser, as it is usually thought. More light might be shed onto instabilities of bifurcation branches by using lower values of η in simulation. Indeed, at $\eta \approx 1$ a very strong force returns any excited atom back to the ground. The less is the η value the easier is to achieve the same shift and the weaker is the returning force, and the less instability should be expected. Unfortunately, the TdE ground population unambiguously depends upon the frequency of emission, and cannot be changed if we need a certain wave length of the emitted light.

Level Laser

In real lasers this obstacle is bypassed by adding additional levels, the third and the forth. Here we will consider only the third lasing level. Having essentially longer lifetime, A*–subsystem atoms at this metastable level feature higher collective resistance against spontaneous emission.

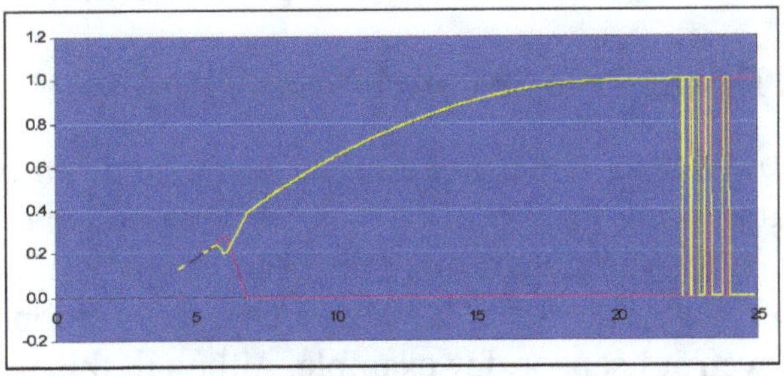

DIBD for map, $\gamma = 0.01$.

That means a strong restriction of their emitting activity which may be linked to reduced thermodynamic activity. Using appropriate activity coefficient γ^*, one may rewrite equation $(1-\eta)/\eta = \exp(-\Delta G^0 / RT)$ as:

$$\Delta G = \Delta G^0 + RT\ln\left[\gamma^* (1-\eta)/\eta\right].$$

In such an approach, the active population of the A^* subsystem equals to $\gamma^* [1-\eta(1-\delta_e)]$, while the ground population still is $\eta(1-\delta_e)$. Now the extended logistic map for the laser states above equation turns to:

$$\ln\left[(1-\eta)/\eta\right] + \ln\left\{\gamma^*\left[1-\eta(1-\delta_e)\right]/\eta(1-\delta_e)\right\} - \alpha\left(1-\delta_e^{\,q}\right) = 0.$$

The activity coefficient in application to the atom emission may be roughly estimated as a ratio of the lifetimes on the uppermost level and meta-stable intermediate level, that is:

$$\gamma^* \approx \tau_{up} / \tau_{meta}.$$

An example of graphical solutions to this map with a quite modest value $\gamma^* = 10^{-2}$ is given in Figure. It is easy to see that even at that, definitely not the least of possible values of γ^*, the occurrence of spontaneous emission is pushed far right on the DIBD, to the area where almost all atoms belong to the A*–subsystem, making the laser functioning more stable and the thermodynamic picture more realistic. Obviously, the thermodynamic allowance is far not the only factor that permits laser to function and delays or promotes spontaneous emission, though thermodynamic prohibition is the strongest of the negative factors.

Laser Beam Characterization

Laser beams often need to be comprehensively characterized in various respects.

Optical Power

For the measurement of the optical power of a laser beam, there are various types of power meters, which may be based on photodiodes or on some kind of thermal detector. For permanent monitoring, there are optical power monitors.

Spatial Aspects: Beam Profiles and Beam Quality

A full characterization of a monochromatic laser beam could in principle lead to the complete complex amplitude profile in one plane perpendicular to the beam, from which the further evolution (propagation) of the beam could be calculated with suitable beam propagation software, for example. Such a characterization can be done with a Shack–Hartmann wavefront sensor, for example. Although this measures field amplitudes and wavefront directions only on some grid of spots, the amplitudes at intermediate points can be interpolated, if the spatial resolution is good enough. Alternative measurement methods may be based on interferometry, for example.

For non-monochromatic laser beams, one often assumes that the amplitude pattern is not frequency-dependent. That assumption may be completely wrong, however; for example, different resonator modes of a solid-state laser generally have different optical frequencies, so that different frequency components of the generated beam can have entirely different beam profiles. For lasers with operation on a single transverse mode, however, the mentioned assumption can be fairly accurately fulfilled.

There are also various kinds of beam profilers which measure only the intensity distribution, but not the optical phase. They can be based on cameras, for example. A comprehensive assessment of beam quality (quantified e.g. with the beam parameter product or the M2 factor) is nevertheless possible, if such intensity profiles are taken at different positions along the laser beam.

Some laser beams exhibit "hot spots", i.e., regions with higher intensity, which may play role in damage phenomena. There can also be satellite structures, halos and other deviations from beam uniformity. Note that the shape of the beam intensity profile may change during propagation, and is also not always stable over time.

Some devices do not measure the whole beam profile, but only a beam radius or diameter. Note that the beam radius is usually not defined as a half width at half maximum (HWHM) value, but as the radial position where the intensity falls to 1/e2 times the peak intensity. This comes from the context of Gaussian beams.

A laser beam may exhibit beam pointing fluctuations, which can be measured by monitoring the position (center of gravity of the intensity distribution) at a suitable location. The polarization state may be roughly characterized based on the power transmission of a polarizer for different orientations, or more completely with various techniques for polarimetry.

Temporal and Spectral Aspects

For pulsed lasers, measurements of the pulse energy, pulse duration and peak power are often required. In case of Q-switched lasers, optical energy meters and fast photodiodes can be suitable. For mode-locked lasers, generating much shorter pulses, one uses instruments like auto-correlators all those for more advanced methods like FROG and SPIDER. The optical spectrum of a laser beam can be measured with some kind of spectrograph or another kind of spectrometer.

Required Optical Attenuation

For various methods of laser beam characterization, for example with cameras, it is necessary to more or less attenuate a laser beam without e.g. affecting its beam profile. Straight-forward approaches like using some absorbing neutral density filters often do not work at high power levels. Therefore, advanced types of optical attenuators may be needed.

Laser Diffraction

Laser diffraction has emerged as one of the most important and effective techniques in the world of particle size analysis thanks to its fast, non-destructive properties, its suitability for a broad range of particle sizes, and its ability to be fully automated. As a technique that measures particle size distribution for both wet and dry dispersions, it offers many advantages, including a high level of precision, fast response, high potential for the repetition of results, and a wide measurable particle diameter range.

The Role of Laser Diffraction in Particle Analysis

Laser diffraction has, to a large extent, replaced traditional methods of particle size analysis, such as sieving and sedimentation. Recognized for its capacity to reproduce results and size range spanning five orders of magnitude, laser diffraction has emerged as the technique of choice throughout the pharmaceutical industry where examining particle size is crucial in determining the performance of a product or process.

One example of this is the efficacy of 'semisolid' drugs that are often used in ointments, creams, gels or lotions. Semisolid drugs have some of the properties of solids and some of the properties of liquids, so understanding the size of the particles they contain is crucial in knowing how each particular product should be delivered to the human body.

The scope for automation means modern particle size analysis can often be a matter of loading the sample and hitting a button, which is an exciting prospect for pharmaceutical companies looking to scale their research.

How does Laser Diffraction Work?

Laser diffraction is grounded in the relationship between light and surfaces. When light and surfaces interact, it results in either solely or a mix of refraction, reflection, absorption or diffraction. The latter offers the greatest scope for accurate particle size analysis assuming the diffraction system contains the following:

- A laser: This is necessary as a source of intense and coherent light that's of a defined wavelength.
- A sample presentation system: This ensures that the material being tested successfully travels through the laser beam as a stream of particles that have a known state of dispersion and can be reproduced.
- Detectors: Specialised detectors (typically an array of photo-sensitive silicon diodes) are applied to measure the light pattern produced across a range of angles.

Laser diffraction is what is known as a 'cloud' or 'ensemble' technique meaning it offers a result for the entire sample, as opposed to providing information for individual particles. Ensemble techniques use a broadened beam of laser light which scatters the light on to a specialised lens to offer a greater collection. During a laser diffraction experiment, particles are illuminated in a collimated laser beam – producing a scattered pattern of light – allowing scientists to deduce particle size and shape.

As a general rule, the bigger particles will bring about a high intensity of scattering at low angles to the beam and the smaller particles, on the other hand, create a low-intensity signal at far wider angles. These angular scattering patterns are measured with various specially-designed detectors and particle size distribution is determined from the resulting data.

Laser Diffraction Models

Laser diffraction relies on optical models to help scientists understand data produced. The Mie scattering theory and the Fraunhofer diffraction approximation are two key theories used to calculate the type of light intensity distribution patterns produced by particles of various sizes.

Fraunhofer Theory

In the late 1970s, when laser diffraction systems were first introduced, limited computing power made it difficult, and impractical, to rigorously apply Mie theory. The Fraunhofer approximation of the Mie theory was a much easier model to use and was therefore widely adopted at this stage. It provides a simpler approach by additionally assuming that:

- The particle size must be relatively large. It is recommended that 10x the laser wavelength is the minimum for this approximation.
- The particles being measured are opaque discs.
- Light is scattered only at narrow angles.
- Particles of all sizes scatter light with the same efficiency.
- The refractive index difference between the particle and the surrounding medium is infinite.

Mie Scattering Theory

Mie theory uses the refractive index difference between the particle and the dispersing medium to predict the intensity of the scattered light. It also describes how the absorption characteristics of the particle affect the amount of light which is transmitted through the particle and either absorbed or refracted. This capability to account for the impact of light refraction within the particle is especially important for particles of less than 50μm in diameter and/or those that are transparent. Mie theory is based on the following assumptions:

- The particles being measured are spherical.
- The suspension is diluted, so that light is scattered by one particle and detected before it interacts with other particles.
- The optical properties of the particles and the medium surrounding them are known.
- The particles are homogeneous.

Solid-State Laser

Solid-state lasers are the original, conventional workhorses of the bench-top world, differing from semiconductor lasers in size, function, and application. Solid-state lasers are lasers whose active medium is typically a minority ion in a solid-state host. The host is most often a single crystal with about 1% of a different species, such as a neodymium ion (Nd^{3+}), doped into the solid matrix of the host. For certain applications the host is a glass, the most important of these being glass doped with the erbium ion (Er^{3+}), which is then pulled into a long thin fiber and used in telecommunications systems. Solid-state lasers use optical pumping by other lasers or lamps to produce a wide variety of laser devices. The population inversion (a condition in which the atomic population of the higher energy state is greater than the lower energy state) in solid-state lasers is created when the active medium absorbs photons from an intense light source.

One important class of solid-state lasers consists of the tunable and ultrafast lasers. These systems use solid-state lasers to excite them, and have other similarities in their thermal and optical properties.

The typical optically pumped, solid-state laser system is much larger than a semiconductor device. The optical pump may either be a single, "high-power" semiconductor diode, a small array of semiconductor diodes on one chip, or a stack of arrays for a laser with outputs of up to 1000 W. Before the advent of laser diodes, solid-state lasers were pumped by lamps either running continuously for modest power lasers, or pulsed flash lamps for laser output energy in the 1-Joule-per-pulse category. Since the number of arrays needed to power a 1-J-per-pulse laser is still quite pricey, lamps still have an important niche in the solid-state laser business.

The active medium of a solid-state laser consists of a passive host crystal and the active ion, and it is these components that give the laser its name. A "neodymium:YAG" (Nd:YAG) laser, for example, consists of a crystal of yttrium aluminum garnet (YAG) with a small amount of neodymium added as an impurity. It is the Nd ion (Nd^{3+} added in the form of Nd_2O_3 to the materials to make the single crystal) in which the population inversion is created, and which generates the photon of laser light. Typically the lasing ion is present at about 0.1% to 1% of the ion density of the metal ions of the host crystal or glass. There are times, however, when the host of the active ion in the solid-state laser is used as the name of the laser. The Nd:YAG is often called a YAG laser, harking back to a time when the only good solid-state laser was YAG with Nd^{3+} ion minority constituent.

Today, however, there are perfectly good solid-state lasers that use YAG as the host for Yb (ytterbium, lasing at 1.03 μm), Ho (holmium, 2.1 μm), Tm (thulium, 2 μm) or Er (erbium, 2.9 μm) ions. In all of these cases the host material is usually configured in a simple rod shape that is cut from a synthetically grown crystal or poured glass. For the important case of the erbium-glass medium, the glass with erbium minority constituent is drawn into a very long thin fiber and typically used as an amplifier in an erbium doped fiber amplifier (EDFA). In general, the vendor of a laser crystal is not the manufacturer of the laser itself.

The first solid-state laser, indeed the first laser ever, was the ruby laser. This laser used a flash lamp to create a population inversion in a cylindrical rod cut from large single crystals of Al_2O_3 containing small amounts of impurities in the form of C_r2O_3. This mixture has the common name of ruby, given to the natural gem material long before its chemical composition (or the concept of a laser) was known. After the ruby laser was demonstrated, other ion/host combinations were investigated.

Nd:YAg Lasers

The Nd:YAG laser is the most prevalent of today's solid-state lasers. These systems can be found on the industrial work floor welding heavy metals, in the surgical suite performing delicate surgery, in the research laboratory making precise spectroscopic measurements, and on a satellite orbiting the planet Mars measuring the detailed topography. The Nd:YAG lasers found in each of these diverse applications have distinct characteristics that make them suitable for that use. The welding laser, for example, is likely to be a large, lamp-pumped laser, while the surgical system is likely to be a smaller, diode-pumped system that is frequency doubled to produce green light. The spectroscopic laser could also be a diode-pumped system, perhaps utilizing special resonators to reduce its bandwidth. And the Mars-orbiting laser would be Q-switched to produce short pulses for distance measuring.

The neodymium ion can be added as an impurity to a glass matrix, producing the ion/host combination of an Nd:glass laser. Although the same triply ionized neodymium ion does the lasing in both Nd:YAG and Nd:glass, the two lasers have little in common. Even the wavelengths of the two lasers are slightly different (because the electric fields in the two hosts shift the energy levels differently). Nd:glass is much better at storing energy than Nd:YAG, so Nd:glass lasers are used in high-energy, Q-switched applications. Large Nd:glass lasers can produce pulses of 100 J and greater, while Nd:YAG is limited to about 1 J from a single, Q-switched oscillator. On the other hand, the thermal conductivity of glass is much lower than that of YAG, so YAG lasers are preferred for high-average-power applications. A single Nd:YAG lamp-pumped oscillator, containing several Nd:YAG rods lined up in series, can produce an output of a kilowatt or more.

Thermal conductivity is the mechanism for removing waste heat from the interior of the laser host rod. A coolant then removes the heat from the rod surface. If the heat isn't removed fast enough from the interior of the crystal, the host overheats and distorts the optical quality of the crystal.

The triply charged erbium ion (Er^{3+}) can be added as a minority to YAG, glass, and other hosts. In YAG, the lasing wavelength of 2.9 µm is "eye safe," meaning it cannot damage the retina. The strong absorption of 3-µm light by tissue makes the Er^{3+}:YAG ideal for medical applications. In glass, Er^{3+} can lase at 1.5 µm, a wavelength of minimal loss in optical fibers. Erbium-doped fiber amplifiers (EDFAs), glass fibers with Er^{3+} doping excited by semiconductor lasers, have been widely deployed in global telecommunication systems, where they directly boost the signal carried on fibe roptic cables. Before EDFAs were integrated into the fiber optic-based phone system the optical signal in the fiber-based cables had to be converted to an electrical signal for amplification, and then back to optical for further transmission. The advent of EDFAs in the mid-1990s has revolutionized wideband telecommunications.

Another ion, one that absorbs the pump light more efficiently than the lasing ion, is sometimes added to these ion/host solid-state lasers. For example, a Ho:YAG laser might be "co-doped" with chromium (Cr^{3+}) or thulium (Tm^{3+}) ions that strongly absorb the pump light and transfer the energy to the lasing holmium ion. The EDFAs rely on energy transfer from Yb ions co-doped into the glass fiber.

Diode-Pumped Solid-State Lasers

Historically, lamp-pumped lasers preceded their diode-pumped brethren by decades, though the simplicity of diode pumping makes the laser easier to explain. Lamp pumping uses a relatively inexpensive light source but is considerably less efficient than diode pumping. Diodes or diode arrays are more expensive per unit of energy but despite the cost of powerful semiconductor diode arrays, diode pumping often is preferable because it produces much less heat in the laser medium and has a significantly greater overall efficiency.

The energy level diagram for Nd^{3+} is a four-level system, meaning that the pump light goes into a state that is not the upper laser level, relying on the energy put in the laser to relax to the upper level in a billionth of a second or less. Similarly, at the bottom of the energy ladder, the lower level relaxes quickly, thus the name four-level laser. These levels aren't the only energy levels in the Nd ion, and some of the other levels can serve as the lower laser level of secondary lasing transitions, at 1338 and 946 nm. One can force the laser to lase at one of the secondary transitions by maximizing the feedback at that wavelength, and minimizing the feedback at competing laser wavelengths.

Many other ion/host combinations can be diode pumped. Neodymium in other hosts behaves similarly to the Nd:YAG case. Holmium lasers, with output in the 2- to 3-μm region, are often co-doped with thulium (Tm). The Tm ions efficiently absorb the pumping radiation from a laser diode at 785 nm, and transfer the energy to the lasing Ho ion. The laser diodes that supply this 785-nm pump lamp are similar to the diodes that pump Nd:YAG at 808 nm. Er:glass lasers can likewise be co-doped with ytterbium (Yb), which absorbs pump light at 960 nm and transfers the energy to the erbium.

But not all ion/host combinations are suitable for diode pumping. Ti:sapphire, for example, requires green pump light, at a wavelength too short to be generated efficiently by existing laser diodes. Moreover the storage time in Ti:sapphire is much shorter than that for Nd:YAG, and a huge number of diodes would be needed to pump energy into the Ti^{3+} ions. As such, Ti:sapphire lasers tend to be laser pumped, with either argon-ion lasers, CW diode-pumped Nd lasers or pulsed, lamp-pumped Nd lasers providing the optical pumping energy.

Thin-Disk Laser

Thin-disk lasers are solid-state bulk lasers having a very thin disk of laser-active material as the gain medium. Power scalability and minimal thermal lensing are the particular features of the thin-disk laser. High output power, high efficiency and good beam quality are further advantageous qualities of this type of solid-state lasers. Since the invention of the thin-disk laser 16 years ago, thin-disk lasers evolved to a common tool for cutting and welding. The highest published output power of an Yb:YAG single disk laser is 5.3 kW in multi-mode operation. Nowadays, thin-disk lasers for material processing with an output power up to 16 kW, using four disks, are commercially available. These lasers operate typically in a high-order multi-mode (M2 ≈ 20).

Interest has been shifted towards a significant enhancement of the brightness. This would make new laser applications at long distances more practicable, especially in the fields of material processing as well as security and defense. However, in the multi-kilowatt regime, thermal lensing is even in thin-disk lasers not negligible. Therefore, a thin-disk laser resonator can be considered as a resonator containing an internal lens. Thermally induced phase distortions of the field distribution in the resonator can reduce the laser efficiency. Furthermore, these aberrations yield a degradation of the beam quality. This aspect is quite irrelevant in multi-mode operation. But in fundamental mode operation, it is necessary to reduce the influence of these aberrations or even to compensate for them, systematically. Dynamically stable resonators are well suited to reduce the influence of internal lenses. Dynamically stable means that the radius of the laser mode in the active medium passes an extremum depending on the refractive power of an internal / thermal lens. Near this extremum, a variation of the refractive power of the internal lens would yield only a very small variation of the mode structure.

Characterization of Thin-Disks

For the purpose of the experimental realization of lasers with high brightness, the design of a dynamically stable resonator requires a thorough characterization of all components. A diagram of the performed characterization of thin-disk modules is displayed in Figure.

Firstly, the refractive power, the aberrations as well as the pump power distribution are investigated. On the basis of these investigations, numerical simulations of the field distribution in the resonator are performed, in order to identify the influences of the aberrations on the beam quality. In combination with thermo-mechanical computations, the optical path differences (OPD) of the disks itself are analyzed, in order to improve alternative thin-disk designs. Based on these studies, the thin-disk laser resonators are designed and concepts for compensating the OPD are developed.

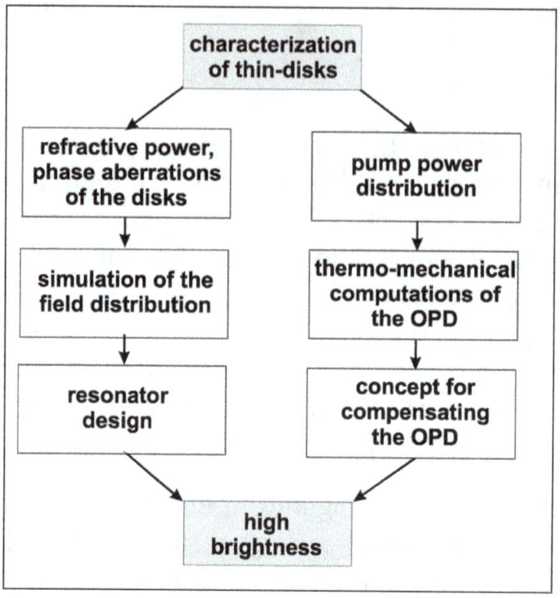

Diagram of the characterization of thin-disks with respect to the enhancement of the brightness of thin-disk lasers.

In order to demonstrate the power scalability of thin-disk lasers, Yb:YAG thin-disk laser modules with large pump spot radii (r_p = 4.15 mm), large disk radii (r_{disk} = 7 mm) and a maximum pump power P_p = 1.8 kW per module were used. The Yb:YAG thin-disks with a doping concentration of 12.5 % and a thickness of 140 µm are bonded on a 1.4 mm thick diamond substrate. The modules itself were manufactured by Trumpf Laser GmbH & Co. KG. The basic principle of thin-disk lasers is displayed in Figure.

In the following, the characterization of thin-disk laser modules is described, including measurements of the refractive power and of optical path differences (OPD) as well as thermo-mechanical computations of the disks.

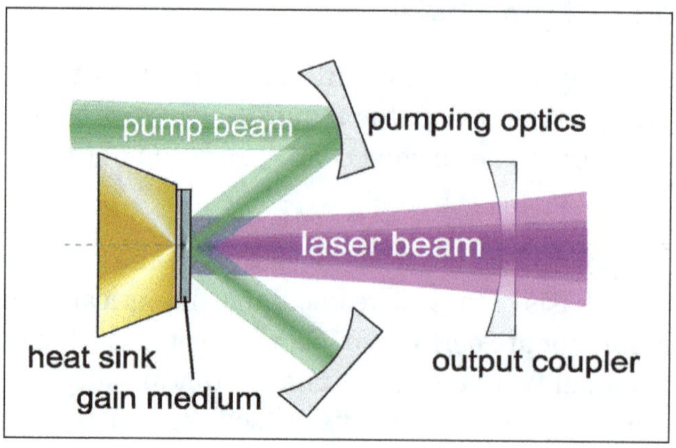

Basic principle of the thin-disk laser: The gain medium is a thin crystalline disk, bounded on a heat sink. The rear side of the disk is HR coated for the laser wavelength as well as for the pump wavelength. The disk itself acts as a HR mirror of the laser resonator as well as a bending mirror for the pump beam. The pumping optics is designed for a multi-pass transit of the pump beam through the disk.

Refractive Power of Thin-Disks

For adapting the stability range of the resonator to the variation of the radius of curvature of the disk, the refractive power of the disks was measured as function of the pump power in multi-mode operation.

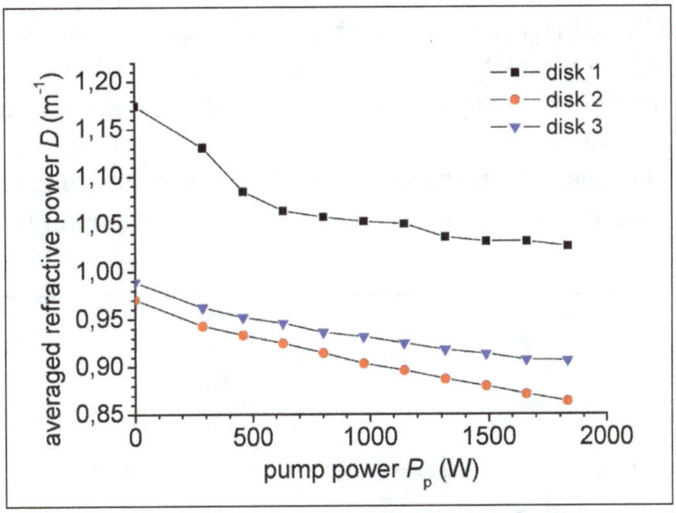

Averaged refractive power D of three exemplary thin-disks as function of the pump power P_p in multi-mode laser operation.

The refractive power D of the disks shows a nearly linear decrease as function of the pump power P_p. This linearity is in accordance with predictions of thermo-mechanical computations.

The typical reduction of the refractive power is 7 - 15 cm^{-1} for a pump power up to 1.8 kW. At maximum pump power, the refractive powers of the disks are between 0.86 m^{-1} and 1.03 m^{-1}. Comparing the refractive powers along the x- and y-axis, there is also a slight astigmatism. The differences of these refractive powers are about 4 % of the maximum value. For designing the resonator, the minimum refractive power, i.e. maximum focal length, of the disks at maximum pump power, was taken as an upper limit of the thermal lens.

Optical Path Difference

The quantum defect in fluorescence processes, stimulated emission as well as amplified spontaneous emission (ASE) processes yield an increase of the temperature of the disk with increasing pump power. The resulting increase of the refractive index, the thermal expansion as well as the deformation of the disk are equivalent to a variation of the optical path difference (OPD) which is important for the field distribution in the resonator. In order to develop concepts for the compensation of phase distortions as well as for an improvement of alternative designs of disk-bonding, OPD measurements have been performed.

The total OPD of a thin-disk can be considered as a sum of a parabolic part and a remaining non-parabolic part. Deviations of the OPD from the ideal parabolic shape cause phase distortions of the field distribution. The parabolic part of the OPD can be simply compensated by a variation of the resonator parameters, e.g. varying the radius of curvature of a mirror or varying the resonator length.

The deviations of the OPD and thus the phase distortions depend very sensitively on the homogeneity of the intensity distribution of the pump radiation. They can yield a reduction of the optical efficiency in fundamental mode operation and a degradation of the beam quality due to an admixture of higher order modes. In order to investigate the deviations of the OPD from the ideal shape, measurements of the OPD with a Shack-Hartmann sensor were performed. The experimental setup is shown in Figure below.

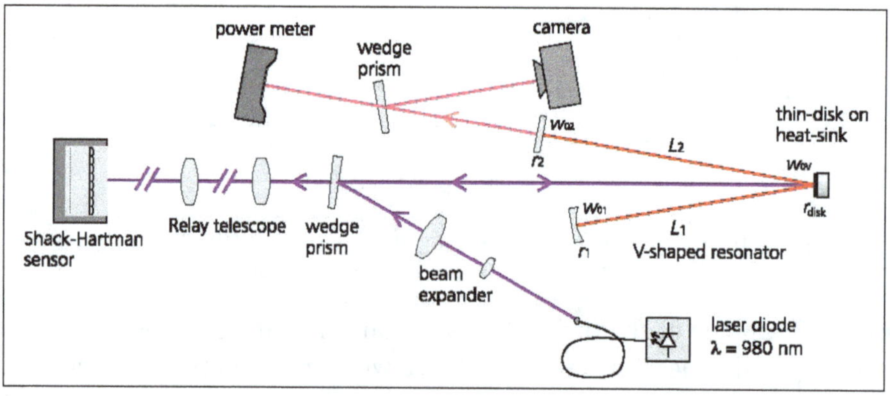

Experimental setup for measurements of the optical path difference (OPD). The investigated disk is irradiated by a collimated probe beam (λ = 980 nm). The wavefront in the plane of the thin-disk is imaged on a Shack-Hartmann wavefront sensor, using a Relay telescope. The beam diagnostics include the detection of the laser output power and the beam profile. The measurements of the OPD were performed in laser operation mode (fundamental or multi-mode operation) as well as in fluorescence operation mode.

The measurements of the OPD were performed in laser operation mode (optional multi-mode or fundamental mode operation) as well as in fluorescence operation mode. The investigated disks show qualitatively similar OPDs in both operation modes.

The OPDs of the investigated thin-disks show deviations from the ideal parabolic shape, depending on the pump power. Besides astigmatism, there is a radial increase of the deviations. The diagrams of the OPD show a good agreement with the subtracted paraboloid within the pump spot (r_p = 4.15 mm). In this area of the disk, the optical path differences are smaller than 100 nm.

To allow for fundamental mode operation, the radius woV of the laser mode on the disk has to be about 70 % of the pump spot radius r_p. Therefore, an important criterion for fundamental mode operation is the degree of the deviations of the OPD within the radius of the laser mode on the disk. For compensating these deviations, the utilization of intra-cavity adaptive optics would be useful. If the spatial resolution of the deformability of an adaptive optical element is high enough, this could be also used to compensate for the non-parabolic part of the OPD.

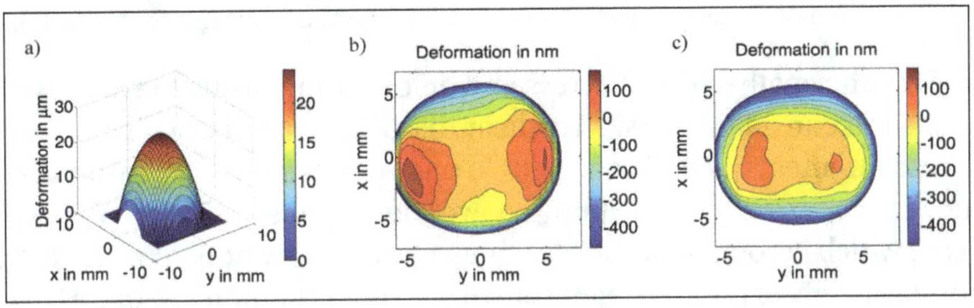

Optical path difference (λ = 980 nm) of a typical disk in fundamental laser operation mode: a) total path difference with P_p = 1030 W; b) with P_p = 0 W after subtraction of an axially symmetric paraboloid z = $(x^2 + y^2)/(2rdisk)$, with r_{disk} = 2.01 m; c) as b), but with P_p = 1030 W, r_{disk} = 2.19 m.

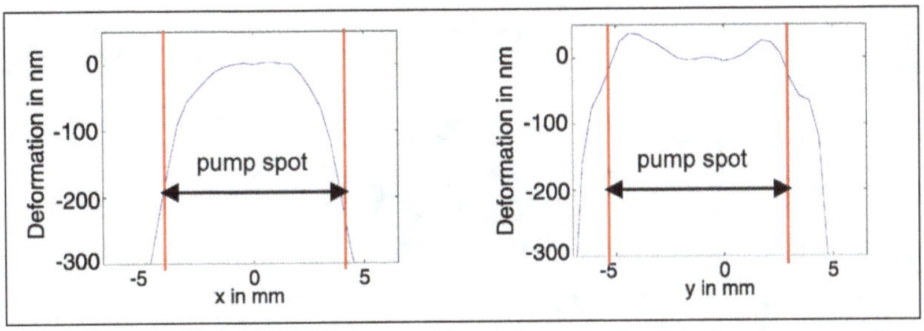

Cross sections of the optical path differences along the x- and y-axis (pump power P_p = 1030 W, pump spot radius r_p = 4.15 mm, r_{disk} = 2.19 m).

Thermo-mechanical modeling of thin-disks For the purpose of further optimization of the design of thin-disks laser modules, thermo-mechanical computations was performed, using a finite elements (FE) software. Especially to discriminate between the influence of inhomogeneous pumping and the influence of non-perfect mounting of the disk, we evaluated these path differences for an idealized design numerically. Figure below shows the assumed mechanical setup: the thin-disk is bonded (e.g. glued) to a disk made of diamond, this diamond disk is hold in an annular copper structure. No initial curvature of the disk was assumed. The diamond is cooled from the bottom. Thermo-mechanical modeling of thin-disks For the purpose of further optimization of the design of thin-disks laser modules, thermo-mechanical computations was performed, using finite elements (FE) software. Especially to discriminate between the influence of inhomogeneous pumping and the influence of non-perfect mounting of the disk, we evaluated these path differences for an idealized design numerically. It shows the assumed mechanical setup: the thin-disk is bonded (e.g. glued) to a disk made of diamond, this diamond disk is hold in an annular copper structure. No initial curvature of the disk was assumed. The diamond is cooled from the bottom.

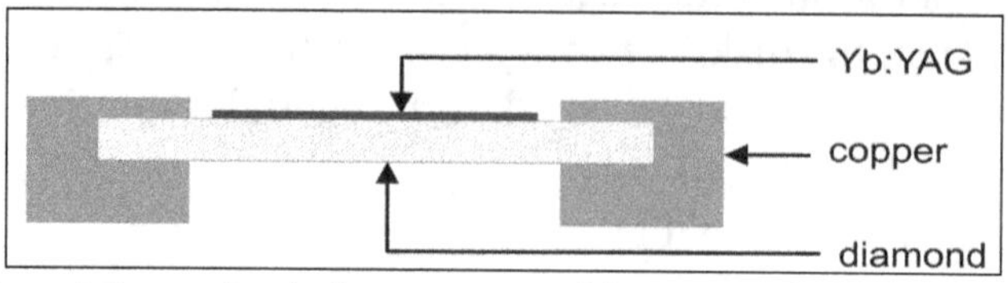

Cross-section of the mechanical structure used for FEM calculations (not to scale).

To calculate the thermally induced stress and deformation inside the disk, we applied a volume heat source inside the disk. The heat distribution was calculated based on fluorescence measurements of a disk in double pass pump set-up. We assumed that this fluorescence is proportional to the pump power density distribution. We used numerical raytracing which accounts for the bleaching of the absorption to calculate the spatial distribution of the absorption in the disk with a 16 pass pump set-up. At 1 kW pump power we calculated an absorption efficiency of about 90 %. With a quantum defect of about 10 %, this generates about 90 W of heat inside the disk. For the distribution of this heat inside the disk, we assumed that the heat is generated proportionally to the previously calculated spatial distribution of the absorption. The resulting distribution of the heat inside the disk is sketched in Figure below.

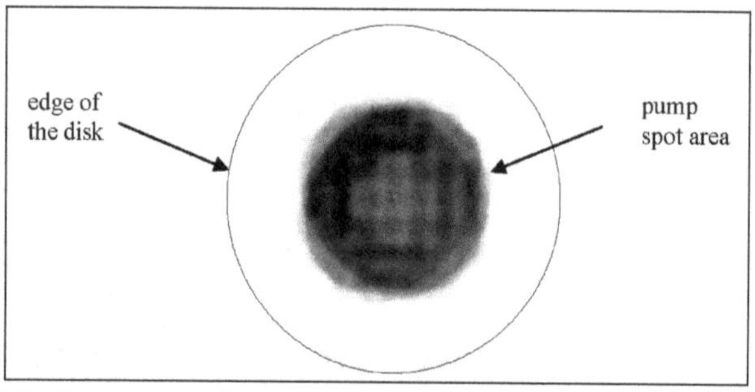

Distribution of the heat source density in the disk, max. Value about 18 kW/cm³, 90 W total heat source (white = low density, black = high density).

From the results of the finite element software, the optical path difference (OPD) Φ (r) can be calculated by:

$$\phi(r)=2\left[\int_0^{h_{disk}}[n_0+\frac{\partial n}{\partial T}(T(r,z)-T_0)+\Delta n_s(r,z)-1].[1+\varepsilon_z](dz)-z_0(r)\right]$$

with n_0 the refractive index at the reference temperature T_0, $\partial n/\partial T$ the thermo-optical coefficient, Δ_{ns} the change of refractive index due to stresses, ε_z the strain in z-direction and z_0 the displacement of the HR side of the disk, h_{disk} the thickness of the disk. A cross section of the resulting optical path difference is shown in Figure.

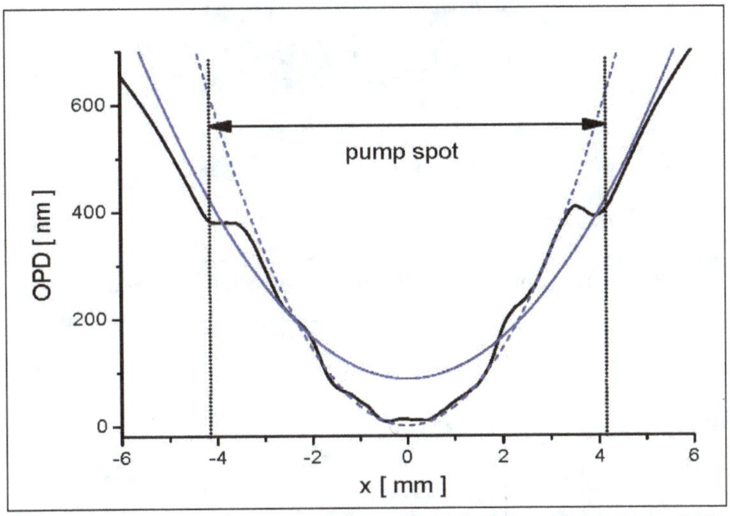

Cross section of the calculated optical path difference (thick line: Total OPD, solid line: paraboloid with D = 3.9 cm^{-1}, dashed line: paraboloid with D = 7.2 cm^{-1}, pump spot radius r_p = 4.15 mm).

As the path difference has a mainly parabolic shape, it is useful to split it into two parts: a parabolic part, which can be expressed as an optical refraction power associated with the disk, and a remaining path difference, the so-called non-parabolic part of the thermal lens. Subtracting a parabolic phase (refractive power D = 3.9 cm^{-1}) yields the resulting non-parabolic OPD shown in Figure below. A cross section of this remaining optical path difference is depicted in Figure, showing a variation of the path difference of about 200 nm across the pump spot. We can also assume a higher optical refraction power, as can be seen in, this will lead to a similar variation across the pump spot (and a stronger variation across the whole disk), but to a significantly reduced path difference inside a radius of 3 mm (less than 50 nm).

In comparison with the measurements, the thermo-mechanical modeling shows similar deviations from the parabolic shape. Depending on the refractive power of the subtracted axially symmetric paraboloid, the deviations of the OPD within the pump spot area of the disk are small (< 100 nm), but radially increasing for $r > r_p$. Subtracting suitable parabolids with smaller refractive power, the deviations of the OPD are in a range of ±100 nm over the whole disk.

The measured astigmatism cannot be explained by the assumed inhomogeneity of the pump power distribution. This may be due to the influence of the bonding or mounting process for real disks. The different results for the thermal lens can be explained by a different stiffness of the heat sink and/or the annular copper structure. Additionally, the adequate refractive power strongly varies with the aperture. Characterizing thin-disks, the variation range of the refractive power as function of the pump power as well as the phase aberrations were determined.

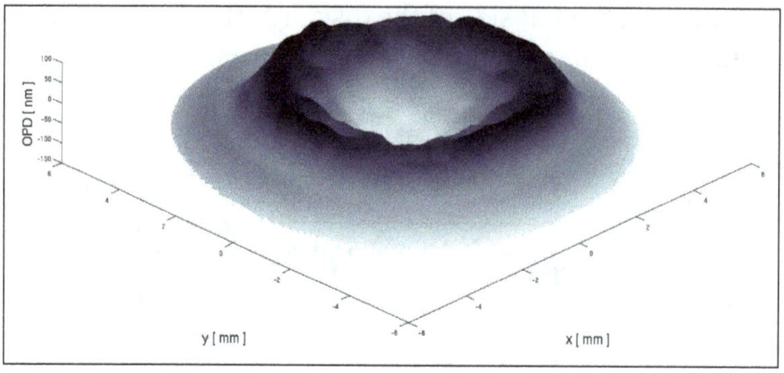

Calculated optical path difference, corrected by 3.9 cm-1.

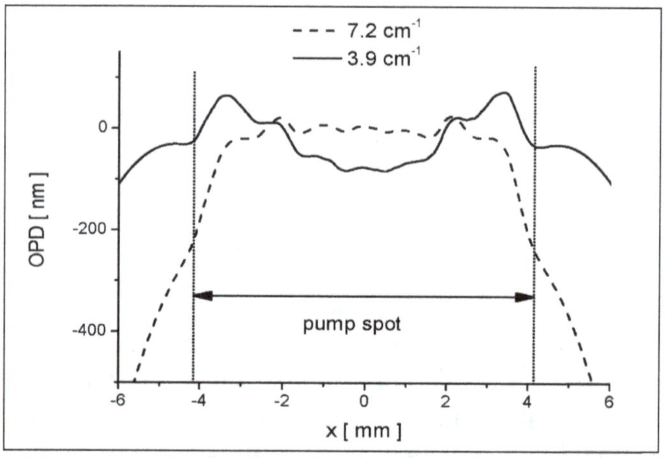

Cross sections of the calculated optical path difference, corrected with different values of the optical refraction power (pump spot radius r_p = 4.15 mm).

Thin-Disk Lasers with Dynamically Stable Resonators

Thermal lensing in thin-disk lasers is not negligible in the high power regime. Therefore, a thin-disk laser resonator can be considered as a resonator containing an internal focusing element. However, for fundamental mode operation, the radius of the laser mode has to be adapted to the radius of the pump spot and the resonator should be almost insensitive against thermal lensing. Dynamically stable resonators can combine these two features, simultaneously. A dynamically stable resonator is a special type of stable resonators. An ordinary dynamically stable resonator contains one internal lens between the two laser mirrors. Usually, dynamically stable resonators have two ranges of stability. Each range has a point of insensitivity of the mode radius to the variation of the refractive power of the internal lens. The development of concepts for the implementation of more than two disks in one resonator is in preparation.

V-Shaped Resonator with One Disk

A simple dynamically stable thin-disk laser resonator is a V-shaped resonator containing one thin-disk module.

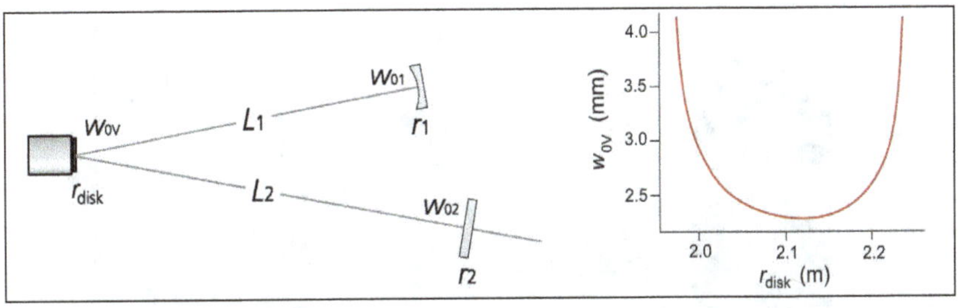

In above figure Left shows Dynamically stable V-shaped resonator with one thin-disk as bending mirror (L_1 = 2.305 m, L_2 = 2.198 m; mirror radii: r_1 = 0.5 m, r_2 =∞, curvature radius of the disk r_{disk} ∈ [2.0 m; 2.22 m]; calculated beam radii of TEM_{00} on the mirrors, on the disk: w_{01} = 0.285 mm, w_{02} = 0.254 mm, w_{0V} = 2.843 mm; pump spot radius r_p = 4.15 mm; reflectivity of the output coupler R = 0.96). Right: Calculated beam radius woV of TEMoo on the disk as function of the curvature radius of the disk r_{disk}.

Due to the reabsorption in the unpumped area of the thin-disk, the pump spot acts as a soft aperture. For small mode radii w_{0V} of TEM_{00}, higher order laser moder can oscillate. By increasing w_{0V} these higher order modes can be suppressed effectively. If w_{0V} gets too large the losses of TEMoo increase too. An optimum for fundamental mode operation is reached for $w_{0V} \approx 0.7_{rp}$.

According to the resonator configuration in Figure, the TEM_{00} mode radius w_{0V} passes a minimum with increasing radius of curvature of the disk r_{disk}. If laser operation is provided near the minimum ($r_{disk} \approx$ 2.12 m), the mode radius on the disk gets almost independent of the pump power. Classically, one would set the operating point at maximum pump power near to this minimum to reduce the influence of the thermal lens. Instead, we used a resonator design where this minimum is reached at medium pump power and the optimum ratio w_{0V}/r_p is reached at maximum pump power.

For the values of the mirror radii of curvature ri as well as the propagations lengths L_i in Figure, the stability range fits very well to the variation range of r_{disk}, i.e. to the variation range of the pump power P_p. For r_{disk} = 2.0 m at P_p = 0 W the TEMoo mode radius woV is about 3 mm = 0.7rp (pump spot radius r_p = 4.15 mm), just like for r_{disk} = 2.2 m at P_p ≈ 1.4 kW. For these values of r_{disk}, fundamental mode operation can be obtained. For values between these two limits of r_{disk}, there is an admixture of higher order modes to the fundamental mode.

The laser emission starts in TEM_{00} mode operation at the threshold, changes continuously to a mixture of higher order laser modes with increasing pump power and then, after w_{0V} passing the minimum value, the mode structure turns back to the fundamental mode. This is confirmed by numerical simulations of the far field as well as the field distribution inside the resonator, considering real pump power distributions on the disk. In this way, we could demonstrate fundamental mode operation of a dynamically stable resonator with one disk, with a maximum output power P_{out} = 496 W (pump power P_p = 1367 W) and with an optical efficiency η = 0.362.

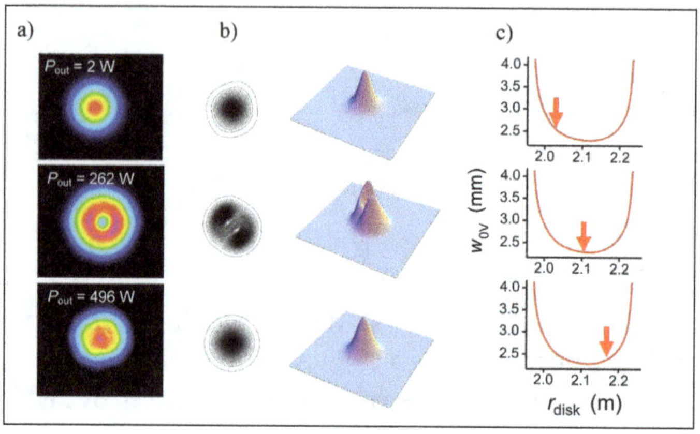

Measured laser mode profiles at P_{out} = 2 W, 262 W and 496 W. b) Contour plots and 3D graphics of the numerical simulations of the field distribution in the far field. c) Radius of TEM_{00} on the disk as function of the radius of curvature r_{disk} of the disk. The arrows indicate the radius of curvature of the disk at the corresponding output power. For r_{disk} = 2.22 m, the optimal ratio of fundamental mode radius to pump spot radius is fulfilled ($w_{oV} / r_p \approx 0.7$).

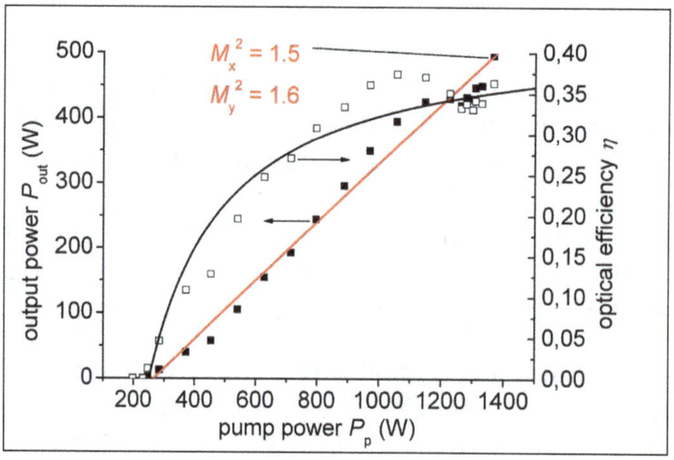

Output power P_{out} and optical efficiency η of the dynamically stable V-shaped resonator with one disk as function of the pump power P_p. The maximum output power in fundamental mode operation was Pout = 496 W with an optical efficiency of η = 0.362 and with $M_x^2 = 1.47$, $M_y^2 = 1.63$ (measured with Data Ray WinCam D).

W-Shaped Resonator with Two Disks

The plane output couplers of two V-shaped resonators are joined to one HR bending mirror, in order to "combine" the two resonators to one resonator.

Experimental setup of a dynamically stable W-shaped resonator with two disks (L_1 = 2.305 m, L_2 = 2.198 m, L_3 = 2.116 m, L_4 = 2.250 m; mirror radii: r_1 = 0.5 m, $r_3 = \propto$, r_5 =

0.5 m; reflectivity of the output coupler R = 0.92; curvature radii of the thin-disks: $r_{disk\ 1} \in [2.00\ m; 2.22\ m]$, $r_{disk\ 2} \in [1.96\ m; 2.16\ m]$; pump spot radii on both disks $r_p = 4.15$ mm).

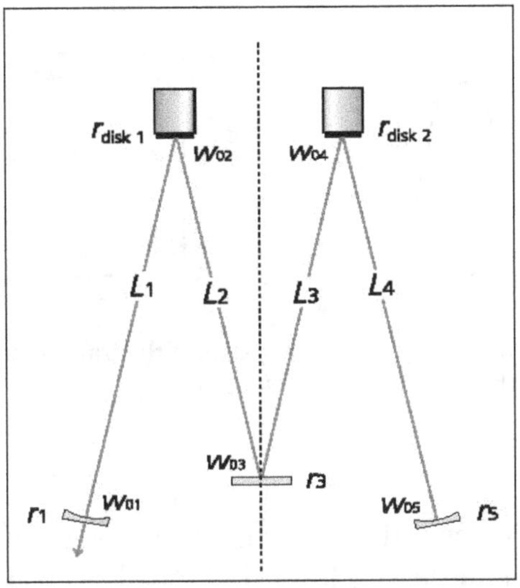

Due to the slight difference of the refractive powers of the two thin-disks, the propagation lengths L_i of each V-shaped part of the W-shaped resonator have to be adjusted in order to get nearly identical mode radii w_{02} and w_{04} on both disks. As mentioned above, for fundamental mode operation a ratio of radii $w_{02,04} / r_p$ should be about 0.7. Using this resonator, we get nearly fundamental mode operation with a maximum output power P_{out} = 904 W with a pump power P_p = 2464 W. Disadvantageously, the heat generation in the disk induces turbulences in air, especially in front of the thin-disk. These turbulences yield fluctuations of the output power up to 5 % of its maximum value. Compared to high order multi-mode operation, there is a considerable reduction of the optical efficiency in fundamental mode operation due to the fluctuations of the refractive index of hot air.

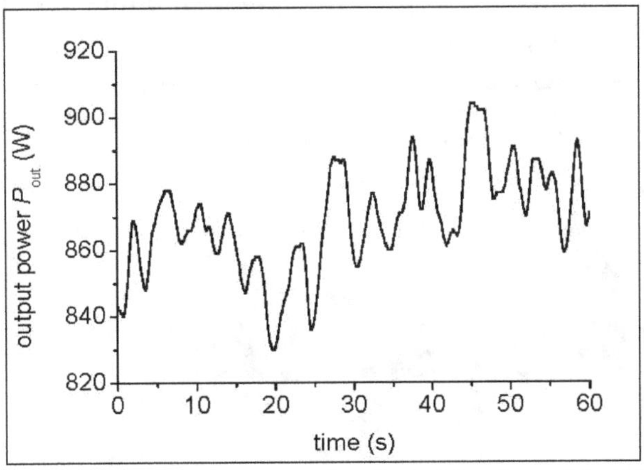

Measured output power Pout as function of the time with slightly increasing pump power up to P_p = 2464 W.

Unfortunately, the turbulences in air were so strong that they inhibited the measurement of M^2 at this pump power. For this reason, we measured the beam profiles at P_{out} = 788 WZ. For an output power Pout = 904 W the beam profiles were similar to these.

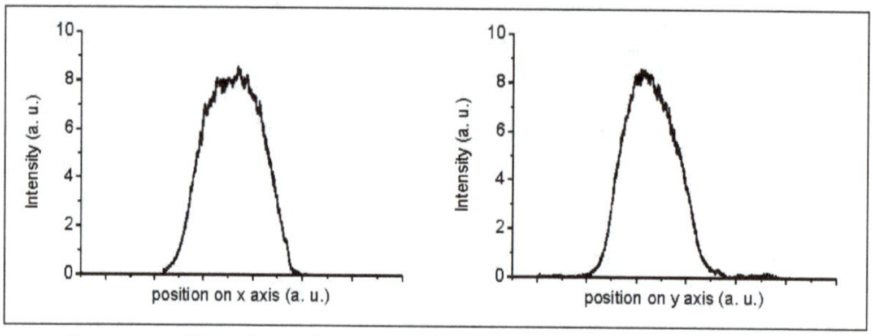

Beam profiles along the x- and y-axis of the transverse plane of the thin-disk laser resonator (P_{out} = 788 W).

Ruby Laser

A ruby laser is a solid-state laser that uses the synthetic ruby crystal as its laser medium. Ruby laser is the first successful laser developed by Maiman in 1960. Ruby laser is one of the few solid-state lasers that produce visible light. It emits deep red light of wavelength 694.3 nm.

Construction of Ruby Laser

A ruby laser consists of three important elements: laser medium, the pump source, and the optical resonator.

Laser Medium or Gain Medium in Ruby Laser

In a ruby laser, a single crystal of ruby ($Al_2O_3 : Cr^{3+}$) in the form of cylinder acts as a laser medium or active medium. The laser medium (ruby) in the ruby laser is made of the host of sapphire (Al_2O_3) which is doped with small amounts of chromium ions (Cr^{3+}). The ruby has good thermal properties.

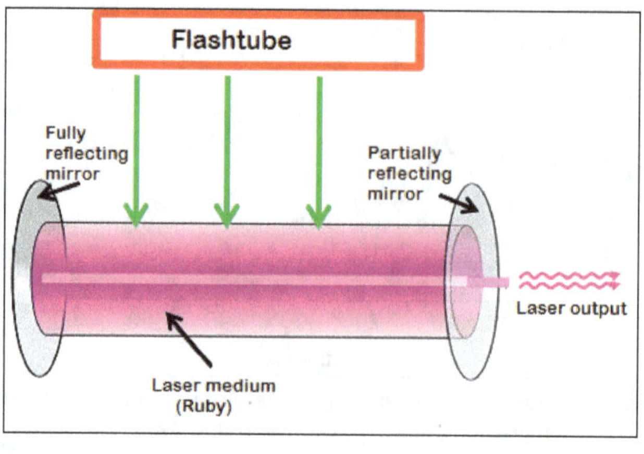

Pump Source or Energy Source in Ruby Laser

The pump source is the element of a ruby laser system that provides energy to the laser medium. In a ruby laser, population inversion is required to achieve laser emission. Population inversion is the process of achieving the greater population of higher energy state than the lower energy state. In order to achieve population inversion, we need to supply energy to the laser medium (ruby).

In a ruby laser, we use flashtube as the energy source or pump source. The flashtube supplies energy to the laser medium (ruby). When lower energy state electrons in the laser medium gain sufficient energy from the flashtube, they jump into the higher energy state or excited state.

Optical Resonator

The ends of the cylindrical ruby rod are flat and parallel. The cylindrical ruby rod is placed between two mirrors. The optical coating is applied to both the mirrors. The process of depositing thin layers of metals on glass substrates to make mirror surfaces is called silvering. Each mirror is coated or silvered differently.

At one end of the rod, the mirror is fully silvered whereas, at another end, the mirror is partially silvered. The fully silvered mirror will completely reflect the light whereas the partially silvered mirror will reflect most part of the light but allows a small portion of light through it to produce output laser light.

Working of Ruby Laser

The ruby laser is a three level solid-state laser. In a ruby laser, optical pumping technique is used to supply energy to the laser medium. Optical pumping is a technique in which light is used as energy source to raise electrons from lower energy level to the higher energy level. Consider a ruby laser medium consisting of three energy levels E_1, E_2, E_3 with N number of electrons.

We assume that the energy levels will be $E_1 < E_2 < E_3$. The energy level E_1 is known as ground state or lower energy state, the energy level E2 is known as metastable state, and the energy level E_3 is known as pump state. Let us assume that initially most of the electrons are in the lower energy state (E_1) and only a tiny number of electrons are in the excited states (E_2 and E_3).

When light energy is supplied to the laser medium (ruby), the electrons in the lower energy state or ground state (E_1) gains enough energy and jumps into the pump state (E_3). The lifetime of pump state E_3 is very small (10^{-8} sec) so the electrons in the pump state do not stay for long period. After a short period, they fall into the metastable state E_2 by releasing radiation less energy. The lifetime of metastable state E_2 is 10^{-3} sec which is much greater than the lifetime of pump state E_3. Therefore, the electrons reach E_2

much faster than they leave E_2. This results in an increase in the number of electrons in the metastable state E_2 and hence population inversion is achieved. After some period, the electrons in the metastable state E_2 falls into the lower energy state E_1 by releasing energy in the form of photons. This is called spontaneous emission of radiation.

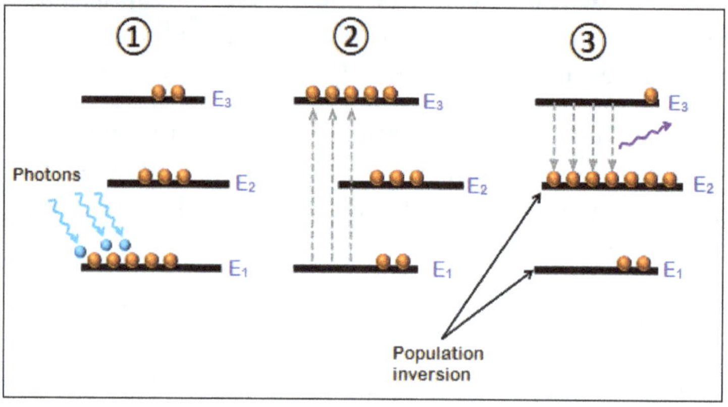

When the emitted photon interacts with the electron in the metastable state, it forcefully makes that electron fall into the ground state E_1. As a result, two photons are emitted. This is called stimulated emission of radiation.

When these emitted photons again interacted with the metastable state electrons, then 4 photons are produced. Because of this continuous interaction with the electrons, millions of photons are produced.

In an active medium (ruby), a process called spontaneous emission produces light. The light produced within the laser medium will bounce back and forth between the two mirrors. This stimulates other electrons to fall into the ground state by releasing light energy. This is called stimulated emission. Likewise, millions of electrons are stimulated to emit light. Thus, the light gain is achieved. The amplified light escapes through the partially reflecting mirror to produce laser light.

Gas Laser

Gas laser is a laser with a gaseous active medium. A tube containing the active gas is placed in an optical resonator, which in the simplest case consists of two parallel mirrors, one of which is semitransparent. A light wave emitted at any point in the tube is amplified upon propagation through the gas by events of induced emission, which give rise to a photon avalanche. Upon reaching the semitransparent mirror, the wave partially passes through it. This part of the light energy is radiated outward by the gas laser. The other part is reflected from the mirror and gives rise to a new photon avalanche. All the photons are identical in frequency, phase, and direction of propagation. Because of this, laser radiation may have extraordinarily high mono chromaticity and power and sharp directivity.

The first gas laser was developed in the USA in 1960 by A. Javan. Existing gas lasers operate in a very broad range of wavelengths—from ultraviolet to far infrared—in both the pulse and continuous modes. Certain data on the most widely used continuous-wave lasers are presented in table below.

Table: Continuous-wave lasers.

	Wavelength (μm)	*Output (watts)*
Cadmium...............	0.3250	a few ten-thousandths
Cadmium...............	0.4416	tenths
Argon...............	0.4880	units
Argon...............	0.5145	tens
Krypton...............	0.5682	units
Helium-neon...............	0.6328	tenths
Helium-neon...............	1.1523	tenths
Xenon...............	2.0261	hundredths
Helium-neon...............	3.3912	hundredths
CO...............	5.6-5.9	hundreds
CO_2...............	9.4-10.6	tens of thousands
HCN-molecular...............	337	thousandths

Of the gas lasers operating only in the pulse mode, lasers in the ultraviolet band operating on neon ions (λ = 0.2358 microns [μm]) and on N2 molecules (λ = 0.3371 μm) are of the greatest interest. The nitrogen laser has a high pulse power.

The characteristic properties of laser radiation—high directivity and mono chromaticity—are manifested most distinctly in the radiation of a gas laser. A significant advantage of gas lasers is their ability to operate in the continuous-wave mode. The use of new methods of excitation and the transition to higher gas pressures can sharply increase the power of gas lasers. By using gas lasers it is possible to make use of the far infrared band, as well as the ultraviolet and X-ray bands. New fields for the use of gas lasers—for example, in space studies—are opening up.

Particular Features of Gases as Laser Materials

In comparison with solids and liquids, gases have a much lower density and higher homogeneity. Therefore a light ray in a gas is virtually not distorted, does not disperse, and experiences no energy losses. In such lasers it is comparatively simple to excite just one type of electromagnetic wave (one mode). As a result, the directivity of the laser radiation is sharply increased, reaching a limit determined by light diffraction. The divergence of the light ray of a gas laser is 10^{-5} to 10^{-4} radians in the visible light band and 10^{-4} to 10^{-3} radians in the infrared region.

In contrast to solids and liquids, the particles comprising a gas (atoms, molecules, or ions) interact with each other only upon collisions in the process of thermal motion. This interaction has little effect on the grouping of particle energy levels. Therefore, the energy spectrum of the gas corresponds to the energy levels of the individual particles. The spectral lines corresponding to transitions of the particles from one energy level to another are broadened only slightly in a gas. The narrowness of the spectral lines in a gas leads to a situation in which few resonator modes are incident upon the line.

Since a gas has virtually no effect on the propagation of radiation in a cavity, the stability of the radiation frequency of a gas laser depends primarily on the immobility of the mirrors and the entire structure of the resonator. This leads to the extremely high stability of the radiation frequency of a gas laser. The frequency ω of gas laser radiation is reproduced to an accuracy of 10^{-11}, and the relative stability of the frequency $\Delta\omega/\omega = 10^{-14}$.

The low density of gases impedes the production of a high concentration of excited particles. Therefore the density of the energy generated in gas lasers is much lower than in solid-state lasers.

Production of an Active Gaseous Medium

The aggregate of the excited particles of a gas (atoms, molecules, or ions) that have population inversion is the active medium of a gas laser. This means that the number of particles "populating" higher energy levels is greater than the number of particles at lower energy levels. Under ordinary conditions of thermal equilibrium, the reverse is true—the population of lower levels is greater than that of higher levels. In case of a population inversion, events of induced photon emission at an energy $hv = \varepsilon_h - \varepsilon_l$ which accompany the forced transition of particles from a higher level εh to a lower level ε_l, predominate over instances of absorption of these photons. As a result of this, the active gas may generate electromagnetic radiation with frequency $v = (\varepsilon^h - \varepsilon^l)/h$ (or with wavelength $\lambda. = c/v$).

One of the specific features of a gas (or of a mixture of gases) is the diversity of the physical processes that lead to its excitation and the creation within it of a population inversion. The excitation of an active medium by the radiation of gas-discharge tubes, which has found extensive application in solid-state and liquid lasers, is not very effective in producing a population inversion in a gas laser, since gases have narrow absorption lines, but the tubes radiate light in a broad range of wavelengths. Hence only a negligible part of the power of the pumping source can be used (the efficiency is low). In the overwhelming majority of gas lasers a population inversion is created in an electrical discharge (gas-discharge lasers). The electrons formed in a discharge excite the gas particles upon collision with them (electron impact), causing them to pass into higher energy levels. If the lifetime of particles at a higher energy level is greater than at a lower level, a stable population inversion is formed in the gas. The excitation of atoms and molecules by electron impact is the best-developed method of producing a population inversion in gases. The method may be used for exciting a gas laser in both the continuous-wave and pulse modes.

Excitation by electron impact may be successfully combined with another excitation mechanism—the transfer of the energy needed to excite the particles of one type from particles of another type upon inelastic collision (resonant excitation transfer). Such transfer is extremely efficient when the energy levels of particles of different types coincide. In these cases the creation of the active medium takes place in two stages: first, the electrons excite the particles of the auxiliary gas; then, in the process of inelastic collisions with the particles of the working gas, these particles transfer their excitation energy to them. As a result of this, the upper laser level is populated. To ensure good energy accumulation, the upper energy level of the gas should have a long intrinsic lifetime. It is according to precisely this scheme that population inversion is accomplished in the helium-neon laser.

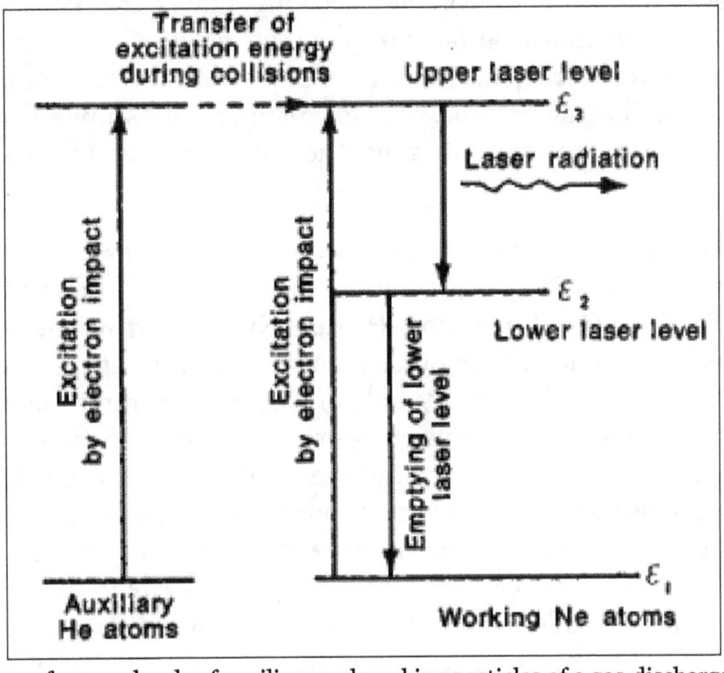

Diagram of energy levels of auxiliary and working particles of a gas-discharge laser.

The Helium-Neon Laser

In a helium-neon laser the neutral atoms of neon (Ne) are the working medium. Atoms of helium (He) serve to transfer the excitation energy. In an electrical discharge part of the Ne atoms pass from the ground level ε_1 to the excited higher energy level ε_3. But in pure Ne the lifetime at level ε_3 is brief, and the atoms quickly "jump" from there to levels ε_1 and ε_2, which hinders the formation of a sufficiently high population inversion for the pair of levels ε_2 and ε_3. An admixture of He substantially alters the situation. The first excited level of He coincides with the upper level ε_3 of Ne. Therefore, upon the collision of H_e atoms (excited by electron impact) and unexcited Ne atoms (with energy ε_1), excitation transfer takes place, as a result of which the Ne atoms will be excited and the He atoms will return to the ground state. When there is a sufficiently large number of a He atom, preferential population of the ε_3 level of neon can be attained. The emptying of the ε_2 level of neon, which takes place during collisions between atoms and the walls of the gas-discharge tube, also contributes to this. For efficient emptying of the ε_2 level, the tube diameter must be sufficiently small; however, a small tube diameter limits the amount of Ne and hence the generated power. From the standpoint of maximum generated power, a diameter of about 7 mm is optimal. Thus, as a result of the special selection of quantities (partial pressures) of Ne and He and upon proper selection of the diameter of the gas-discharge tube, a stationary population inversion of energy levels ε_2 and ε_3 of neon is established.

The ε_2 and ε_3 levels of neon are of complex structure, that is, they consist of many sublevels. As a result, a helium-neon laser may operate at 30 wavelengths in the visible light band and the infrared region. The mirrors of the optical resonator have multilayer dielectric coverings. This makes it possible to create the necessary reflection factor for a given wavelength and thus to excite generation at the required frequency in the gas laser.

The main structural component of a helium-neon laser is the gas-discharge tube, which is usually made of quartz. The gas pressure in the discharge is 1 mm of mercury (mm Hg), and there is usually ten times more He than Number Error. The design of a helium-neon laser developed for use in outer space is shown in Figure. A discharge tube with an internal diameter of 1.5 mm, made of corundum ceramic, is placed between the semitransparent mirror and a reflecting prism that are mounted on a rigid beryllium tube (cylinder). Discharge is accomplished with direct current (8 milliamperes, 1,000 volts) in two sections (each 127 mm long) with a common central cathode. A cold oxide-tantalum cathode (44 mm in diameter and 51 mm long) is divided in half by a dielectric separator, which ensures more uniform distribution of the current over the cathode's surface. Stainless-steel vacuum bellows, which are the anodes, form a movable connection between each pipe and the retainers of the mirror and prism. The housing has an outlet at the left side. The laser is designed to operate in space for 10,000 hours.

The radiation power of helium-neon lasers may reach tenths of a watt, and the efficiency does not exceed 0.01 percent, but the high monochromaticity and directivity of the radiation, the simplicity of handling, and the reliability of design have caused them to be widely used. A red helium-neon laser (λ. = 0.6328 μm) is used in alignment and leveling (mining operations, shipbuilding, and the construction of large structures). The helium-neon laser is widely used in optical communications and location, in holography, and in quantum gyroscopes.

The Carbon-Dioxide Laser

Molecules, unlike atoms, have not only electron but also vibrational energy levels, which result from oscillations of the atoms constituting the molecule with respect to the equilibrium positions. The transitions between vibrational energy levels correspond to infrared radiation. Lasers in which these transitions are utilized are called molecular lasers. Among the molecular lasers, the CO_2 laser, which utilizes the vibrational levels of CO_2 molecules between which a population inversion is formed, is of particular interest.

In gas-discharge CO_2 lasers the population inversion is also accomplished by exciting the molecules by electron impact and resonant excitation transfer. Molecules of nitrogen (N_2), which are in turn excited by electron impact, serve to transfer the excitation energy. Under conditions of a glow discharge, about 90 percent of the nitrogen molecules usually pass into the excited state, whose lifetime is very great. Molecular nitrogen accumulates excitation energy well and transfers it easily to the CO_2 molecules in the course of inelastic collisions. High population inversion is accomplished by adding He to the discharge mixture, which first facilitates conditions for the discharge and second, because of its high thermal conductivity, cools the discharge and contributes to the emptying of the lower laser levels of the CO_2 molecule. Efficient excitation of CO_2 lasers may be accomplished by chemical or gas-dynamics methods.

The fine structure of the vibrational levels of the CO_2 molecule makes it possible to change the wavelength (to re-tune the laser) in increments of 30-50 hertz (Hz) in the wavelength interval between 9.4 and 10.6 μm.

Carbon-dioxide lasers have great power (the maximum power of laser radiation occurs in the continuous-wave mode) and high efficiency. Upon excitation of CO_2 molecules by electron impact and with a gas-discharge tube 200 m long, a CO_2 laser radiates a power of 9 kilowatts (kW). Compact designs with a 1-kW output also exist. In addition to high output, CO_2 lasers have high efficiency, reaching 15-20 percent (efficiency of 40 percent may be attained). In principle CO_2 lasers can also operate efficiently in the pulse mode. These features of CO_2 lasers account for the diversity of their application—in industrial processes (cutting and welding), location and communications (the atmosphere is transparent to waves with $\lambda = 10$ μm), physics research connected with the production and study of high-temperature plasma (high radiation power), and the study of materials.

The gas-discharge tubes of CO_2 lasers have a diameter of 2 to 10 cm, and their length may be very great. Modular (sectional) designs with a discharge current of up to several amperes at voltages of up to 10 kV per section are usually used. Since the power of continuous-wave CO_2 lasers attains very high values, the manufacture of sufficiently long-lived mirrors of good optical quality is a serious problem. Gold-plated sapphire or metal mirrors are used. The radiation is often tapped through apertures in the mirrors. Plastics made of high-resistance germanium, gallium arsenide, or other materials are used as the semitransparent discharge mirrors.

Undesirable effects that destroy the population inversion —heating of the gas and dissociation of its molecules— take place in the electrical discharge of CO_2 lasers. To eliminate them the gas mixture is continuously "driven" through the discharge tubes of the lasers, thus renewing the active medium. In order to produce high power (several kW) in the continuous mode, the gas is driven through the pipe at high speed, and the discharge takes place in a supersonic flow. To avoid losses of costly He, the gas mixture is circulated through a closed loop. Electron impact excitation is accomplished either in the resonator or immediately prior to the mixture's entry into the resonator. In the best devices virtually all CO_2 molecules that enter the resonator are already excited and during their passage through the resonator yield their excitation energy in the form of a quantum of radiation.

Ion Lasers

In ion lasers the population inversion is created between the electron energy levels of the ionized atoms of the inert gases and metal vapor. Population inversion is accomplished by selecting a pair of levels for which the lower level has a shorter lifetime and the upper level a longer lifetime. The need to produce a large number of ions leads to gas-discharge current densities in ion lasers of as high as tens of thousands of amperes per square centimeter. The electrical discharge is effected in fine capillaries up to 5 mm in diameter. At higher current densities the gas is diverted by the current from the anode to the cathode. To compensate for this effect the anode and cathode regions of the discharge tube are joined by an additional long tube of small diameter, which ensures the reverse flow of the gas.

Because of the high current density, cermet designs or pipes made of beryllium ceramics, which have high thermal conductivity, are used to manufacture the gas-discharge tubes for ion lasers. The efficiency of ion lasers does not exceed 0.01 percent. In the visible-light region argon lasers have comparatively high power in the continuous mode. An argon-ion laser generates a radiation with $\lambda = 0.5145$ μm (a green beam) with a power of up to several dozen watts. It is used in the processing of solid materials, in physics research, in optical communications lines, and in optical tracking of artificial earth satellites.

An ion laser using a mixture of argon and krypton ions may, be tuned (by replacing the mirrors) to wavelengths throughout the entire visible spectrum. It radiates a power of up to 0.1 W at wavelengths of 0.4880 μm (blue), 0.5145 μm (green), 0.5682 μm (yellow), and 0.6471 μm (red).

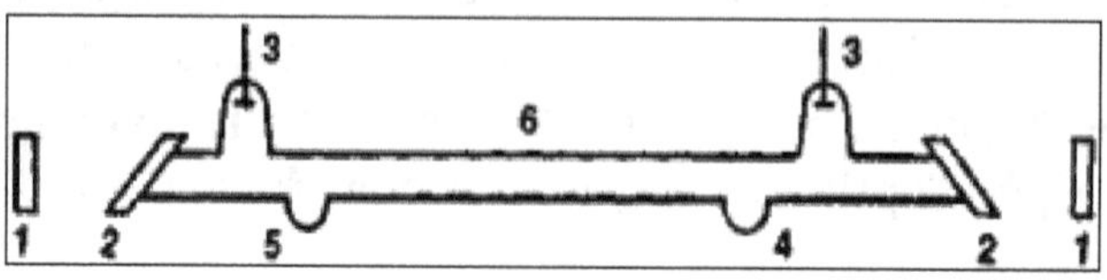

The cadmium-vapor laser, which operates in continuous mode in the blue (0.4416 μm) and ultraviolet (0.3250 μm) regions of the spectrum and has high monochromaticity, is extremely promising. Cadmium vapors are formed in the evaporator placed along the anode. They are strongly diluted with He. Uniform distribution of cadmium in the gas discharge tube and the selection of cadmium concentrations are attained by deflecting cadmium vapors from the anode to the cathode by means of helium ions. The density of the cadmium vapors is determined by the temperature of the heater. The cadmium condenses in the coolant around the cathode. At a helium pressure of 4.5 mm Hg, a heater temperature of 250 °C, a discharge current of 0.12 amps, and a voltage of 4 kV, a pipe 2.5 mm in diameter and 140 cm long makes it possible to produce 0.1 W in the blue and 0.004 W in the ultraviolet regions of the spectrum. The cadmium laser is used in optical research, oceanography, photobiology, and photochemistry.

Gas-Dynamic Lasers

A characteristic feature of gases is the possibility of creating high-speed flows of gaseous masses. If a gas that has been strongly preheated is abruptly expanded—for example, when it passes through a nozzle at supersonic velocity—its temperature drops sharply. When the temperature of a molecular gas drops suddenly, the vibrational energy levels of the molecules may be excited (gas-dynamic excitation). A CO_2 laser using gas-dynamic excitation exists. During gas-dynamic excitation the thermal energy is directly transformed into the energy of electromagnetic radiation. The radiation power of gas-dynamic lasers operating in the continuous mode is as high as 100 kW.

Chemical Lasers

In some gases population inversion may result from chemical reactions in which excited atoms, radicals, or molecules are formed. A gaseous medium is convenient for chemical excitation, since the reacting substances intersperse easily and quickly and are easily transported. Chemical lasers are interesting in that direct conversion of chemical energy into the energy of electromagnetic radiation takes place within them. The excitation that occurs during the chain reaction of the combination of fluorine with deuterium, which results in the production of excited fluorine deuteride (DF), which subsequently transfers its excitation energy to the CO_2 molecules, may serve as an example of chemical excitation. The removal of reaction products ensures the continuous operation of these lasers.

Gas lasers in which population inversion is accomplished with the aid of photo-dissociation reactions (in which the molecules decay when exposed to light) border on the chemical lasers. These are high-speed reactions during which excited radicals or atoms appear. Dissociation takes place under the effect of the radiation of a xenon flashtube. The excited atomic ion I^+ is the reaction fragment.

Dye Laser

A dye laser is a laser based on a dye (typically in a liquid solution) as the gain medium. Most laser dyes are based on organic molecules used in liquid form as solutions, although solid laser dyes and vapor dyes exist. A wide range of emission wavelengths from the ultraviolet to the near-infrared region can be addressed with different laser dyes. Typical characteristics of laser dyes as gain media are:

- They exhibit a broad gain bandwidth, which allows for broad wavelength tunability and also ultrashort pulse generation with passive mode locking.
- Upper-state lifetimes are typically a couple of nanoseconds, i.e., similar to those of semiconductor lasers, but orders of magnitude shorter than those of doped-insulator lasers. This means that dyes are not suitable for Q switching with continuous or long-pulse pumping; they can store energy only over very short times. However, intense dye laser pulses can be obtained with pulsed pumping, using a Q-switched pump laser or a flash lamp.
- The gain per unit length can be fairly high (order of 10^3 cm^{-1}), particularly for pulsed pumping.
- The power conversion efficiency is typically between 10% and 30% for laser pumping, and possibly lower for flash lamp pumping.

Many different dyes are nowadays available, and in total they can cover huge wavelength regions. One name like exalite, coumarin, rhodamine, pyrromethene, pyridine, fluorescein, pyridine or styryl can characterize a whole family of dyes with slightly different chemical structures, having somewhat different ranges of emission wavelengths and being distinguished with additional numbers. For example, one may use coumarin 2, 47, 102 or 153 for lasers in the blue to green spectral region.

Typical solvents for laser dyes are ethanol, p-dioxane and dimethylsulfoxide (DMSO). The used dye concentrations depend on the used pumping geometry and pump power; they are normally well below a gram per liter. The dyes can be purchased in powder form, such that the user can make the dye solution with any desired concentration.

Note that many laser dyes and some of the used solvents are poisonous and partly also carcinogenic. One should therefore carefully avoid to expose the skin to such a dye solution, or even to spill the dye by operating a not properly assembled dye circulating system. A particularly hazardous solvent, sometimes used for cyanide dyes, is dimethylsulfoxide (DMSO), which greatly accelerates the transport of dyes into the skin.

Pumping Options for Dye Lasers

Dye lasers are normally pumped at relatively short wavelengths. Some typical options are:

- Many dyes can be pumped with a green laser such as an argon ion laser or a frequency-doubled solid-state laser. For 532-nm pumping, for example, laser emission may then be in the green, yellow, orange, red or near infrared spectral region.
- Laser dyes for shorter emission wavelengths (for example, exalite, stilbene and coumarin) have to be pumped at shorter wavelengths – typically, with ultraviolet light. For example one may use a Q-switched and frequency-tripled Nd:YAG laser at 355 nm, or a frequency-quadrupled Nd:YAG laser at 266 nm.
- Other possibilities are to use an excimer laser, a nitrogen laser, or a copper vapor laser.
- Some dye lasers can be pumped with flash lamps. This leads to longer pulses and lower power conversion efficiency, but possibly very high pulse energies.

Pulsed pumping allows the excitation of large volumes and thus the generation of pulses with high energies (sometimes of the order of 1 J). Even for continuous pumping, the applied pump intensity is fairly high; for example, a few watts of pump power are focused to a beam waist with a radius of the order of 20 μm.

Dye Laser Resonators

The laser resonator of a dye laser typically contains a dye jet or a dye cuvette as the laser gain medium. In addition, there are usually one or several frequency-selective elements, for example a birefringent tuner or a diffraction grating in Littrow configuration, which permits wavelength tuning in a range of tens of nanometers. Narrow-line width lasers contain additional frequency filtering components such as etalons, and often sophisticated computer-controlled tuning mechanics for precisely setting the wanted emission wavelengths. On the other hand, mode-locked lasers use only course wavelength control, as their emission bandwidth must be large. A stable linear polarization of the output is usually obtained as a side effect of the used tuning elements.

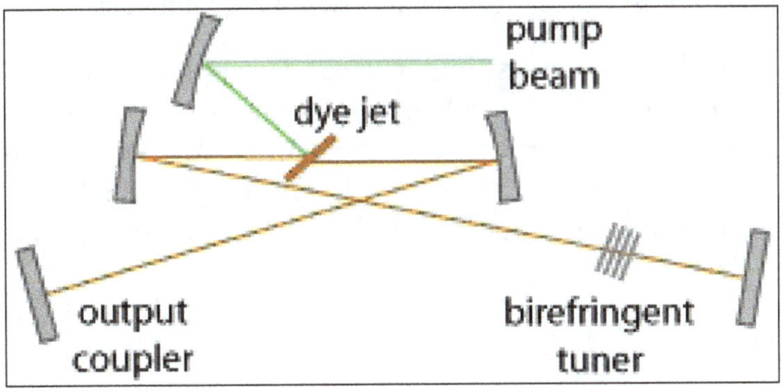

Setup of a wavelength-tunable dye laser.

Dye Jets and Dye Cuvettes: Recovery from Triplet States

A frequently used kind of dye laser uses a thin dye jet as the gain medium, so that the dye molecules are used only for a short time within the pump and laser beam and have a long time to recover before they are used again.

Alternatively, the dye can be pumped through a thin cuvette. Here, the dye is enclosed in some transparent material, and it is easier to obtain a steady flow. Obviously, the cuvette material (for example, quartz) must be resistant to the laser light and the pump light, and the cuvette surfaces must have a high optical quality.

Some time for recovery is often required due to the tendency of organic dye molecules to become trapped in triplet states, in which they cannot participate in the lasing process. An alternative way to lower the triplet concentration is to add a triplet quenching agent to the dye solution.

Degradation: Exchanging Dyes

During operation, laser dyes tend to be chemically degraded. Therefore, the dye solution usually has to be exchanged after an operation time of 100 hours, for example. That lifetime depends on the dye used, on the pumping conditions, the volume of dye solution, etc.

One also sometimes has to exchange the dye in order to access a different wavelength region. There are systems where different dye cells (cuvettes) can be used for different dyes, and switching between different dye cells is relatively easy, requiring little or no realignment of the laser resonator.

Ultrashort Pulse Generation

Much of the original work on ultrashort pulse generation was done with dye lasers. However, dye lasers suffer from significant disadvantages such as rapid degradation during operation, limited output power, and the need for pumping e.g. with green or blue light, making their pump sources expensive. Furthermore, dye lasers require the awkward handling of poisonous, often even carcinogenic materials.

For such reasons, solid-state lasers, in particular Ti:sapphire lasers and later on also fiber lasers, took most of the business from dye lasers (at least in the domain of ultrashort pulse generation) as soon as they were sufficiently developed. Dye lasers are still used in some areas, e.g. laser spectroscopy with wavelengths which are otherwise hard to generate. They are also particularly suitable for intracavity laser absorption spectroscopy.

Solid Laser Dyes

There is some work on dye lasers based on solid media, e.g. with the dye in a polymer matrix. Obviously, the solid-state form has many advantages, particularly concerning handling. A problem, however, is the rapid degradation, either for a limited time by triplet excitation, or permanently by destruction of dye molecules. Some success has recently been achieved with rotating dye disk lasers.

Semiconductor Laser

Semiconductor lasers make powerful, precise beams of light (like ordinary lasers), but they're about the same size as simple LEDs—the little colored lamps you see on electronic instrument panels.

Essentially, an LED is a semiconductor sandwich with the "bread" made from slices of two different kinds of treated silicon known as p-type (rich in "holes" or, in other words, slightly lacking electrons, the tiny negatively charged particles inside atoms) and n-type (with slightly too many electrons). Put the two slices together and you make what's called a p-n junction diode that has all kinds of interesting properties.

In an ordinary diode, the p-n junction works like a turnstile that allows electric current to flow in only one direction (known as forward-biased operation). As electrons flow across this barrier, they combine with holes on the other side and give out energy in the form of phonons (sound vibrations) that disappear into the silicon crystal. In an LED, much the same process takes place but the energy is given out not as phonons but as photons—packets of visible light.

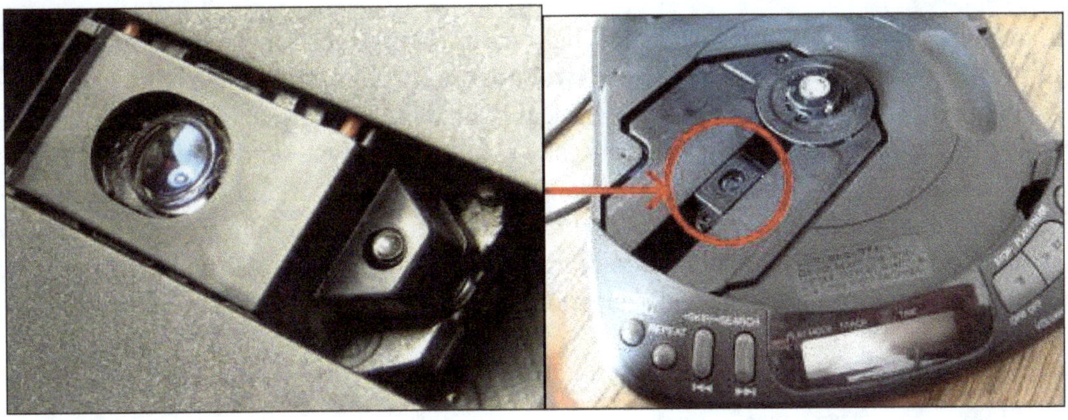

The smaller circle on the bottom left of this photo is a semiconductor laser diode in a CD player. The larger, blue-tinted circle on the top right is a lens that reads the reflected light bouncing down off the CD. Never attempt to look at the laser light in a CD player: you could easily blind yourself.

How Diode Lasers Make Light

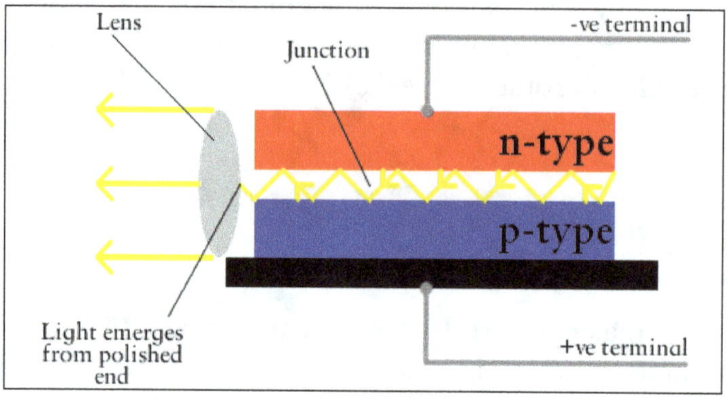

The basic setup of a laser diode. Laser light is produced when electrons and photons interact in a p-n junction arranged in a similar way to a conventional junction diode or LED. One end of the diode is polished so the laser light can emerge from it. The other ends are left roughened to help confine the light.

In a laser diode, we take things a stage further to make the emerging light more pure and powerful. Instead of using silicon as the semiconductor, we use a different material, notably an alloy of aluminum and gallium arsenide (indium gallium arsenide phosphide is another popular choice). Electrons are injected into the diode, they combine with holes, and some of their excess energy is converted into photons, which interact with more incoming electrons, helping to produce more photons—and so on in a kind of self-perpetuating process called resonance. This repeated conversion of incoming electrons into outgoing photons is analogous to the process of stimulated emission that occurs in a conventional, gas-based laser.

In a conventional laser, a concentrated light beam is produced by "pumping" the light emitted from atoms repeatedly between two mirrors. In a laser diode, an equivalent process happens when the photons bounce back and forth in the microscopic junction (roughly one micrometer wide) between the slices of p-type and n-type semiconductor, which is technically known as a Fabry-Perot resonant cavity (a kind of interferometer). The amplified laser light eventually emerges from the polished end of the gap in a beam parallel to the junction. From there, it goes on to read music from your CD, scan the price on your cornflakes, print out your college dissertation, or do a thousand other useful things.

Stacked Laser Diodes

Early laser diodes could fire out only a single, relatively puny beam, but ingenious electronics engineers soon found ways to make them considerably more powerful. Since the 1990s, one common approach has been to mount a number of laser diodes on top of one another (like an apartment building) and then focus their individual beams into a single output beam using a collimator and/or lens. This kind of arrangement is variously called a semiconductor laser stack, stacked laser diode, or just a diode stack. Apart from making more power than a single laser diode, a stack opens up the possibility of generating multiple different wavelengths at the same time (because each laser in the stack can make a different one). Instead of a single P-N junction, there are multiple ones, and the laser light beams emerge from the active layers in between them; typically, there's also at least one tunnel junction between the stacked layers. A single pair of terminals (sometimes called Ohmic contacts) feeds electrical power to the entire stack.

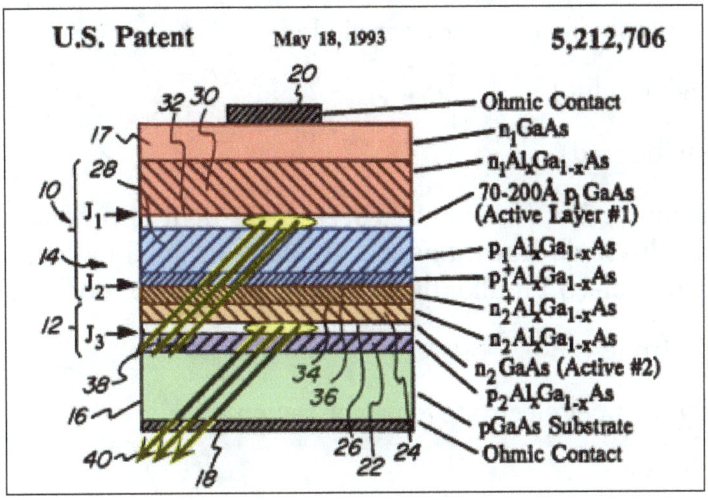

A simple stacked laser diode, comprising two diode lasers one on top of the other, and made from doped layers of aluminum gallium arsenide. The terminals (Ohmic contacts) are shown in gray at the top and bottom, the substrate (base material) is green, P-type layers are shown in blue, and N-type layers in red.

Who Invented Semiconductor Laser Diodes?

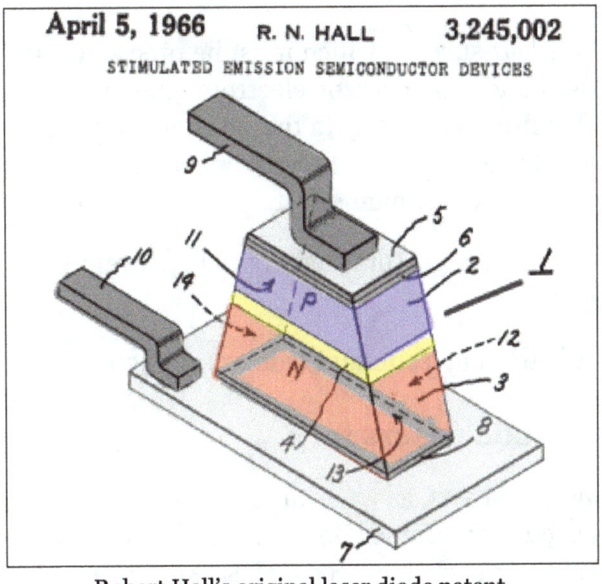

Robert Hall's original laser diode patent.

Who do we thank for this fantastic invention? General Electric's Dr Robert N. Hall, who filed his patent for the idea ("Stimulated emission semiconductor devices") on October 24, 1962 (it was granted as US Patent #3,245,002 on April 5, 1966). Here's one of the drawings from that patent, showing the basic arrangement of the parts described up above. The numbering is Hall's original, but I've added the coloring and simplified descriptions to make it easier to follow:

- A typical P-N semiconductor laser diode.
- P-type region (blue).
- N-type region (red).
- P-N junction region (resonant cavity) where light is produced by stimulated emission. This isn't drawn to scale! In Hall's original patent, it's described as being 0.1 micron (0.1 millionths of a meter, 0.1μm, or 1000 Angstroms) thick.
- Upper electrode.
- Solder fixing upper electrode to p-type region.
- Lower electrode.
- Solder fixing lower electrode to n-type region. (This covers the whole of the base of the n-type region, not just the gray outer outline shown here.)
- Upper electrode connector.
- Lower electrode connector.
- Highly polished front surface.
- Highly polished rear surface, which must be precisely parallel to the front surface to ensure standing waves of electromagnetic radiation (laser light) are produced and emitted efficiently in the resonant cavity between the p-type and n-type regions. Surfaces 11 and 12 may be covered with mirrors or a metallic coating to improve the resonant effect.
- Side surface cut at an angle (or roughened) to prevent waves of light forming in other directions.
- Other side surface cut at a similar angle or roughened in a similar way.

What can we use Diode Lasers for?

Ordinary lab lasers are big beasts, as we've already seen—not so different from the one Goldfinger famously used in the James Bond film of the same name. Putting it another way, anything remotely compact that needs a laser to power it is likely to use a diode laser rather than a "Goldfinger laser." CD players, barcode scanners, fiber-optic phone lines, dental tools, laser hair-removal devices, laser pointers, and laser printers all use diode lasers because they're small, compact, and inexpensive. It doesn't follow that they're low-powered and puny, however—for three reasons.

- Diode lasers can be surprisingly powerful (hundreds of watts is a perfectly achievable output).
- As we saw up above, you can stack diode lasers to make devices with far higher output (in the kilowatts).
- You can team diode lasers up with solid-state lasers (to make what are called diode-pumped solid-state lasers) or with conventional lasers, using them as "optical pumps" (instead of traditional flash tubes) to excite things like ruby rods (giving output in the megawatts).

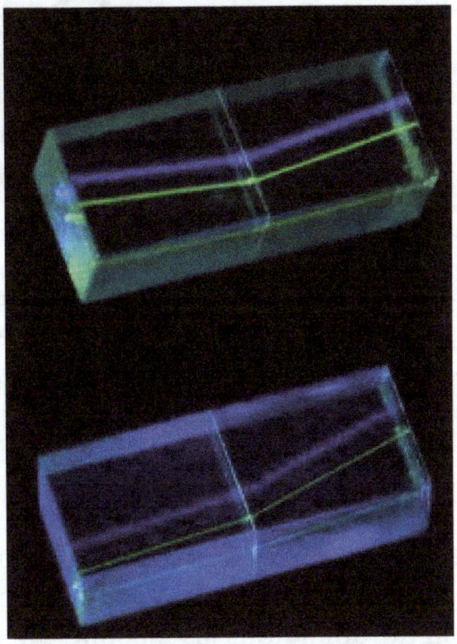

Laser beams bending (refracting) through a crystal.

Common Commercial Varieties

Single-Spatial-Mode Lasers

The simplest and most common laser diode is the single-spatial-mode (also called single transverse mode) laser. Single spatial mode implies that the beam can be focused to a diffraction-limited spot. The term single mode can be, and often is used to refer to both types of laser.

Figure shows a typical single-spatial-mode laser with a relatively narrow waveguide bounded by the cleaved facets. The narrow waveguide, which may be formed by a ridge or by a buried index step, supports only a single optical mode. This limits it to a width of 2 to 5 μm, depending on the details of the waveguide structure. The light emitted by a single-mode laser is, at the facet, typically 1 μm high by 3 to 4 μm wide. Upon leaving the facet, the light diffracts, creating a diverging beam that is typically 20° to 30° (FWHM) in the vertical direction (perpendicular to the plane of the junction, also called the fast axis) and 5° to 10° in the horizontal direction (in the plane of the junction, also called the slow axis).

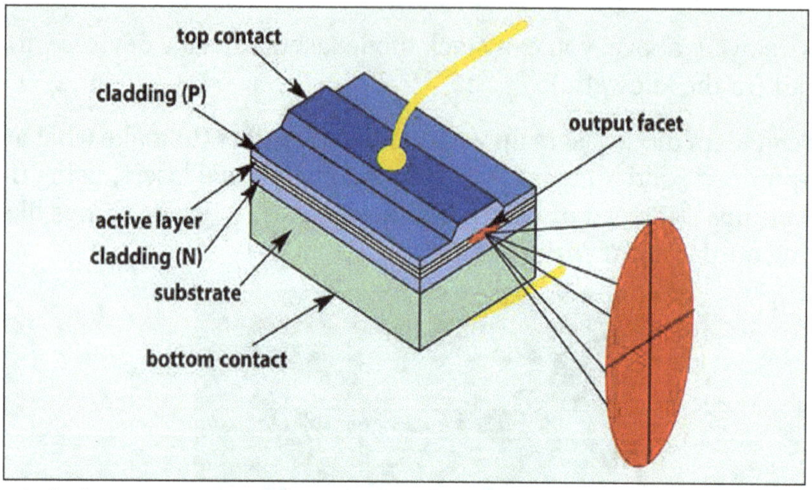

A single-transverse-mode laser diode exhibits the basic laser configuration.

To the optical designer, the beam appears to be an ideal point source. It is, however, anamorphic because the divergences in the two perpendicular directions are different. In addition, many single-spatial-mode lasers are slightly astigmatic, with values ranging from a fraction of a micron up to several tens of microns, depending on the specific laser structure. Both anamorphism and astigmatism can be corrected with relatively simple optics.

Single-spatial-mode lasers are used wherever a diffraction-limited spot is required, such as optical data storage (the largest unit volume application for all laser diodes), thermal and xerographic printing, as pumps and transmission sources in single-mode optical fiber communications and in the ubiquitous laser pointer. Because of the narrow waveguide imposed by the requirement of single-mode operation, the local power density and thermal dissipation are significant and limit the maximum average power to somewhere below 1000 mW. This power can be raised by more sophisticated laser structures.

The Fabry-Perot resonator in a single-spatial-mode laser diode can support multiple longitudinal modes — wavelengths separated by about 1 Å. While many Fabry-Perot diodes can at times run in a single longitudinal mode, they are susceptible to unpredictable mode hops, and either modulation or fairly low levels of feedback can destabilize the laser, causing it to run in multiple longitudinal modes.

Multimode Lasers

The limit on total output power from a single-mode laser comes from considerations of power density within the narrow waveguide. Lower power density and higher powers can be obtained simply by making the laser structure wider, raising the width from 3 or 4 µm up to 50, 100 or 200 µm. The wider lasers can operate at much higher powers – up to 2 W is typical from a 100-µm aperture though higher powers are becoming commercially available (4 to 6 W) – at the expense of losing spatial coherence. These lasers are called broad-area, or multimode lasers.

They cannot be focused to a diffraction-limited spot or efficiently coupled into a single-mode fiber. Note, too, that although the facet may be 30 times wider, the maximum power of a multimode laser is less than 30 times higher than that of a single-mode laser. Maximum power typically increases sub-linearly with laser width.

Nevertheless, many applications such as pumping solid-state lasers and external-drum thermal printing do not require a diffraction-limited spot, and multimode lasers, with their high average powers, are the laser of choice in such applications. Total electrical-to-optical conversion efficiencies can exceed 60 percent in such devices.

Bars and Stacks

If the application requires powers in excess of a few watts, the output of multiple laser emitters must be combined. The next level up in the hierarchy of laser power is the laser bar, which is an array of 10 to 50 side-by-side multimode lasers integrated into a single chip. Standard bar dimensions are either 1.0 cm wide by 500 to 2000 µm long. A single bar can emit 20 to 60 W of average power when operated continuous wave (CW), or over 100 W of peak power when operated in a pulsed mode in which duty cycles can run up to 20 percent.

The most common application for bars is pumping solid-state lasers; commercially available wavelengths, therefore, are those required by various laser crystals. The most common is 808 nm, used for pumping Nd:YAG; others include 785, 792, 915 and 940 nm.

Still higher powers are attainable by mounting bars into stacks, in which all of the bars are driven in series. The bar stack offers high power densities, while the series connection provides a more tractable impedance to the laser drive electronics. A 40-W bar will require 40 to 50 A of drive current at about 2-V drive voltage; multibar stacks increase the drive voltage but have the same current requirements as a single bar.

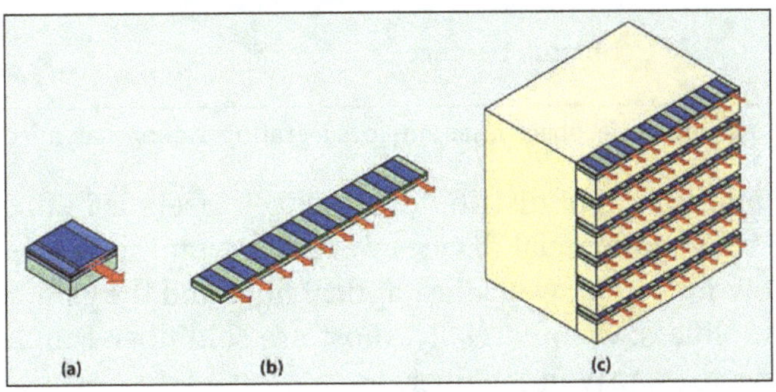

Diode powers (a) can be increased by combining them into bars (b), and stacks of bars (c).

When stacks are used to pump slab lasers, they are not lensed, but simply positioned directly next to the laser crystal. However, for some applications — end-pumping, laser beacons/illuminators and the growing field of direct diode laser materials processing — a stack will be lensed by positioning an array of cylindrical lenses (monolithic or individual, depending on the required brightness) in front of the array of laser bars. At powers above a few tens of watts, the power dissipation exceeds the power handling capability of passive or thermoelectric coolers, and large laser stacks are nearly always water-cooled.

Single-Mode DFB

The ever-growing capacity in fiber optic transmissions systems requires a laser source with low noise and extremely narrow frequency spectrum. This need is met by the distributed-feedback (DFB) laser, which differs from all of the lasers. Instead of relying on mirrors to provide feedback within the laser resonator, it uses a grating — a corrugated structure — embedded within the laser structure in close proximity to the waveguide. Each ripple in the grating provides a minute amount of feedback. The device operates only at the single frequency for which the feedback from the ripples add up in phase. Thus, distributed-feedback lasers operate in a single frequency and remain single frequency under modulation and variations in drive current or operating temperature. Single-frequency operation also eliminates a major source of noise in semiconductor lasers, called partition noise, which results from competition between longitudinal modes in multi-frequency lasers, and therefore they tend to be among the quietest of diode lasers, with relative intensity noise levels of –150 dB/Hz.

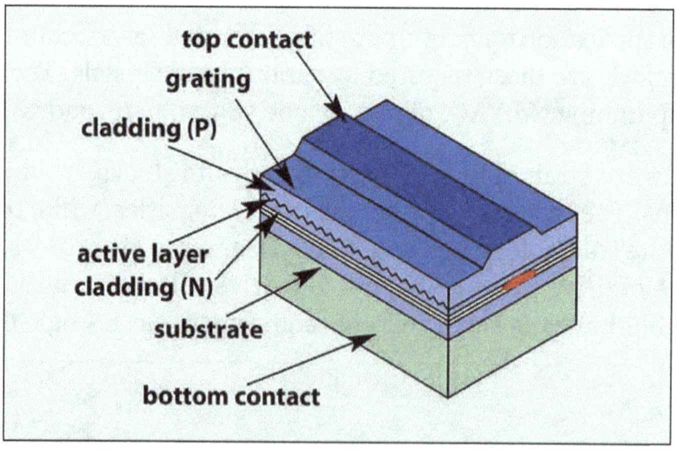

The distributed-feedback laser employs a grating, rather than mirrors.

By far the largest application of distributed-feedback lasers is in fiber optic transmission systems. The first commercial devices were at 1.3 μm; now most are around 1.55 μm to match the low-loss window of silica optical fiber and the gain bandwidth of erbium-doped fiber amplifiers. Commercially, most are sold fiber-coupled; typical output powers range from 1 to 35 mW in the fiber.

A small number are used for various interferometric or spectroscopic applications. In such applications, the desired wavelength is set by a particular atomic or molecular absorption line, or in the case of interferometry, for compatibility with existing gas lasers (e.g., HeNe). Distributed-feedback lasers have been demonstrated for such applications at wavelengths from 633 to 2000 nm and beyond.

Uncommon Laser Structures

Although the majority of lasers sold today fall into one of the above categories, several other types have been demonstrated and are sold commercially. The most novel is the vertical-cavity surface-emitting laser, or VCSEL. Unlike the other lasers presented above, in which the light propagates in the plane of the semiconductor wafer, in the VCSEL, light travels perpendicularly to the wafer and is emitted from the surface. The mirrors are typically provided by growing multilayer dielectric stacks above and below the quantum well active layer; lateral confinement is provided either by gain-guiding or, in newer devices, by an annular ring of oxidized AlAs.

VCSELs are much more difficult epitaxial structures to grow than edge-emitting lasers; the multilayer stacks must have reflectivities as high as 99.8 percent and the layer thicknesses must be precisely controlled to ensure that the resonator frequency falls within the gain bandwidth of the quantum well. However, because they have no cleaved facets, they can be completely fabricated and tested at the wafer level, so their processing subsequent to growth is considerably simpler. In addition, their low power requirements — thresholds can be well below 1 mA — and their low-divergence, nearly circular beam profile make them well suited to coupling to optical fibers. Because of these properties, VCSELs have found a solid application in low-cost, multimode fiber links, particularly in linear and two-dimensional arrays.

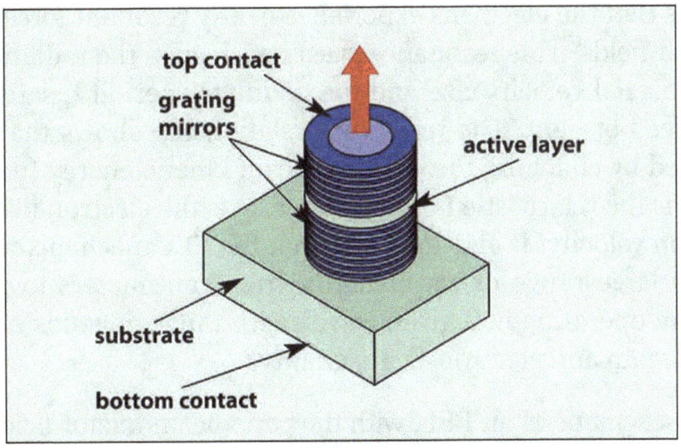

The VCSEL emits in a vertical rather than a horizontal plane.

DBRs and Tunable DFBs

The presence of a buried grating is what makes a distributed-feedback laser operate in a single frequency. More complicated laser structures combine gratings with grating-free waveguide regions and two or more contacts to give monolithic tunable lasers. The distributed Bragg reflector (DBR) laser replaces one of the facets of a Fabry-Perot laser with a grating similar to that of the DFB. It provides higher CW powers than an equivalent DFB laser, at the cost of introducing mode hops with drive current. Addition of a second contact to the grating region gives wideband tunability. DBR lasers and their variants are used for various spectroscopic and instrumentation applications.

Free Electron Laser

The free electron laser (FEL) is a natural alternative as a radiation source. In 1971, J.M.J. Madey invented and developed the FEL, a relativistic electron tube that made use of the open optical resonator. An important step in FEL development came in 1976 when Madey and his co-workers at Stanford University measured gain from an FEL configured as an amplifier at 10-μm wavelength. This experiment, and the successful operation of the same FEL configured as an oscillator in 1977 at 3-μm wavelength, created a large interest in FEL research. Two important FEL attributes, tunability and design flexibility, were demonstrated by these two experiments at significantly different wavelengths using the same apparatus.

Principles of Operation

In an FEL a beam of relativistic electrons produced by an electron accelerator passes through a transverse, periodic magnetic field produced by a magnet called an undulator and exchanges energy with an electromagnetic radiation field. Efficient energy exchange requires that the electrons experience nearly resonant forces from the radiation and undulator fields. This resonance is achieved when the radiation wavelength λ, the electron-beam axial velocity $c\beta z$, and the undulator period λ_o satisfy $\lambda \approx \lambda_o(\lambda - \beta_z)$, where c is the speed of light. The resonance relationship shows that the FEL can be continuously tuned by changing the electron-beam kinetic energy $(\gamma - 1)mc^2$, where $\gamma = (\lambda - \beta_\perp{}^2 - \beta_z{}^2)^{-1/2}$ is the relativistic Lorentz factor, m is the electron mass, and $c\beta_\perp$ is the transverse electron velocity. It also indicates that the FEL mechanism can be designed to operate over a large range of wavelengths, from centimeters to nanometers. The actual feasibility of operating in a given wavelength range depends on several factors, including the FEL gain and electron-beam quality.

Figure below is a schematic of an FEL, with the periodic undulator field indicated by the alternating up and down arrows. The typical peak magnetic field in the undulator is a few kilogauss with a wavelength of λ o $\approx$ several centimeters over N $\approx$ 100 periods. The length of the undulator is several meters inside the optical resonator mirrors, which are separated by about twice the undulator length, or about 10 meters. The electron-beam path is indicated horizontally. The beam energy can vary from a few MeV to several GeV, although the typical FEL uses an electron-beam energy of around 50 MeV. The transverse dimensions in the FEL schematic are exaggerated with respect to the longitudinal distance shown. The electron-beam radius is typically about 1 mm, and the optical mode waist is about 3 mm. The electron-beam pulse width can be as long as several microseconds for a recirculating electrostatic accelerator. In a radio frequency (RF) accelerator or storage ring, the electron-beam structure is determined by the RF frequency. In an RF accelerator, only those electrons with the correct phase relation in the RF cycle continue to be accelerated. The resulting accelerated electron current is produced in short micropulses of a few picoseconds. The minimum separation between micropulses is one RF cycle, but often there is a much larger separation of many nanoseconds. The stream of micropulses continues for about 10 μs up to about 1 ms because the RF power source is pulsed. This stream of micropulses is called a macropulse. In a storage ring using lower-frequency RF, the micropulse length is typically a nanosecond in length separated by several RF cycles. There is no macropulse in a storage ring. The radiation pulse structure in an FEL reflects that of the electron beam.

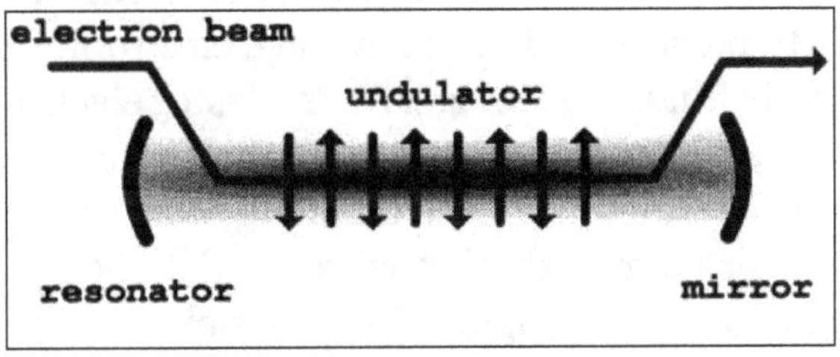

Schematic of a free electron laser.

The requirements of high electron-beam current and quality are crucial in determining the ultimate performance of an FEL. The electron-beam kinetic energy $(\gamma - 1)mc^2$ can range from a few MeV to several GeV, while the beam current can range from a few amperes to several kiloamperes. The FEL is able to obtain large peak power, on the order of a few gigawatts, because, unlike conventional lasers or microwave tubes, the energy that is not transferred to the radiation field remains in the relativistic electron beam that is transported out of the undulator at nearly the speed of light, or even recovered to improve overall efficiency. The FEL interaction volume contains only light, the undulator magnetic field, and the electron beam, so that unwanted high field effects and thermal distortion of the medium are absent in the FEL. The small duty cycle of most accelerators limits the average power of FELs. The efficiency of FELs has been demonstrated to be greater than 40% at long wavelengths, but most will operate at a lower efficiency of a few percent.

The classical gain in the low-gain FEL oscillator develops from coherent electron bunching on the scale of the radiation wavelength. As the electrons travel through the undulator, they accelerate from side to side and spontaneously radiate in the forward direction. The spontaneous emission process is the same as is used to generate UV and x-rays in the undulators and wigglers of synchrotron radiation sources. On the first few passes through the FEL oscillator, some of this spontaneous radiation is stored in the resonator. As the radiation power grows over many passes, the emission process begins to differ from that of the synchrotron sources and becomes similar to the mechanism used in the early microwave electron tubes.

The electrons entering the operating FEL are forced to move from side to side due to the undulator magnetic field, but now radiate in the presence of the stored transverse radiation field. As the radiation passes over the more slowly moving electrons, the radiation electric field does work on the oscillating electrons. This is the classical analog of quantum stimulated emission. A significant energy exchange can occur only when the undulator and radiation field forces on the relativistic electrons are nearly resonant. The electrons in the beam are initially randomly spread over many radiation wavelengths with many millions of electrons within each section of the beam one wavelength long. The orientation of the radiation's electric field vector and the electron's undulator velocity vector rotates through 2π within each radiation wavelength. These different orientations determine that about half the electrons lose energy to the radiation field, while other electrons gain energy from the co-moving wave. The electrons that gain energy begin to move ahead of the average electron, while the electrons that lose energy begin to fall behind the average. As the faster electrons move ahead of the average and the slower electrons move back, the beam is periodically bunched on the radiation wavelength scale. Since the electron energy loss and gain repeats throughout the electron beam over many radiation wavelengths, the beam becomes bunched on the scale of the radiation wavelength. The emission rate for a perfectly bunched beam of electrons is proportional to the square of the number of electrons, whereas the emission rate for a beam of randomly positioned electrons is only proportional to the number of electrons. The same kind of electron bunching is the goal of the microwave electron tube design. The FEL works in a similar way, but can reach infrared, optical, UV, or even x-ray wavelengths. The bunched beam then radiates coherently and amplifies the existing radiation.

There is always an energy spread imposed on the electron beam in order to move the electrons together in a bunch. When the beam has an initial energy, or velocity spread, the bunching process can be impaired, reducing gain. In the FEL oscillator working in steady state at high power, the electrons become over-bunched at the end of the undulator and actually take energy back from the radiation field. This is the normal saturation process, and typically limits the fraction of the electron-beam power converted to radiation in a single pass to 1/N.

In an FEL amplifier (as opposed to an oscillator), the high current of the electron beam increases the rate of the electron bunching process along the undulator. For sufficiently high beam current entering a sufficiently long undulator, the same electron bunching process that occurs over many passes in the FEL oscillator can be completed in a single pass along the FEL amplifier. The amplified wavelength can be selected by an external oscillator laser. In the super radiant FEL, there is no external oscillator, so that radiation must first grow from spontaneous emission. As the optical power grows, the spectrum narrows to establish coherence. This process is called self-amplified spontaneous emission (SASE).

The relativistic electrons "see" each rapidly advancing undulator period Lorentz contracted to a shorter wavelength, $\lambda_{o'} = \lambda_o/\gamma$. Also, the electrons "see" the radiation field passing over them as Doppler shifted to longer wavelengths, $\lambda' = (1 + \beta_z)\,\gamma\lambda \approx 2\,\gamma\lambda$. The condition of resonance between the undulator and radiation forces in the beam frame, $\lambda_{o'} = \lambda'$, gives the FEL resonance condition in the laboratory frame, $\lambda = \lambda_o/2\,\gamma 2$.

Properties

Many of the properties of the FEL can be inferred and these properties are mentioned below:

- Tunability: Because the FEL uses a single gain medium, the relativistic electron beam, and because the resonant condition can be easily tuned by changing either the electron beam energy or the magnetic field strength, FELs are broadly and easily tuned. A factor-of-10 tunable frequency range has already been demonstrated with the same accelerator and undulator.
- High peak power: Because waste energy is carried away at nearly the speed of light and because the lasing medium cannot be damaged by high optical fields, FELs can produce very high peak powers. Giga watt peak powers have been demonstrated.
- Flexible pulse structure: Because the pulse structure of the radiation follows the pulse structure of the electron beam, the mature RF technology of linear accelerators can be used to manipulate and control the FEL pulse structure. Picosecond pulses with sub-picosecond jitter can be produced, the interval between pulses can be varied, and there is the possibility of producing complicated pulse structures.
- Good laser characteristics: FELs easily achieve desirable properties associated with conventional lasers, such as a single transverse mode, high spatial and temporal coherence, and flexible polarization properties.
- Broad wavelength coverage: Because the gain medium is transparent at all wavelengths, FELs in principle can produce radiation at any wavelength. In practice, electron-beam energy, current, emittance, and energy spread requirements become more stringent as the wavelength decreases, and the cost, size, and complexity of the FEL are therefore higher at shorter wavelengths. FELs can have significant emission at harmonics of the fundamental frequency given by the resonance condition, and harmonics can be used to extend operation to shorter wavelengths than would be practical using only the fundamental frequency. At present, the shortest wavelength that has been achieved in an FEL is 240 nm, and significant use of FELs for scientific research has been restricted to the infrared. There are proposals to build vacuum ultraviolet and x-ray FELs.
- Size and cost: Because it requires an electron accelerator with its associated shielding, the FEL has not been a device that can be placed in an individual

investigator's laboratory and be operated and maintained by graduate students whose primary expertise is in other areas of science. FELs have been used principally in central facilities, where their utilization in scientific research involves associated costs of maintaining and operating the facility in addition to the cost of the device itself. FELs have been used by one user, or at most a few users, at any one time, while synchrotrons can be used simultaneously by many users. For example, the APS now under construction at Argonne National Laboratory can have 34 insertion device beam lines and 35 bending magnet beam lines, all of which can be used independently by different users at the same time. The basic science of electron accelerators is well known, but virtually all the effort spent on accelerator physics has been to reach higher energies with bigger and more costly machines. Relatively little effort has been devoted to producing the smaller and less expensive machines that would be most useful for an FEL.

Fiber Laser

The fiber laser is a variation on the standard solid-state laser, with the medium being a clad fiber rather than a rod, a slab, or a disk. Laser light is emitted by a dopant in the central core of the fiber, and the core structure can range from simple to fairly complex. A key factor for fiber lasers is that the fiber has a large surface-to-volume ratio so that heat can be dissipated relatively easily.

Fiber lasers are optically pumped, most commonly with laser diodes but in a few cases with other fiber lasers. The optics used in these systems are usually fiber components, with most or all of the components fiber-coupled to one another. In some cases, bulk optics are used, and sometimes an internal fiber-coupling system is combined with external bulk optics.

A diode pump source can be a single diode, an array, or many separate pump diodes, each with a fiber going into a coupler. The doped fiber has a cavity mirror on each end; in practice, these are fiber Bragg gratings, which can be fabricated within the fiber. There are no bulk optics on the end, unless the output beam goes into something other than a fiber. The fiber can be coiled, so the laser cavity can be many meters long if desired.

Dual-Core Structure

The structure of the fiber used in fiber lasers is important. The most common geometry is a dual-core structure. An undoped outer core (sometimes called an inner cladding) collects the pump light and guides it along the fiber. Stimulated emission generated in the fiber passes through the inner core, which often is single mode. The inner core contains the dopant (ytterbium or erbium) that is stimulated to emit radiation by the pump light. Numerous noncircular variations exist on the shape of the outer core; these shapes, which include hexagonal, D-shaped, and rectangular, decrease the chances of the pump light missing the central core.

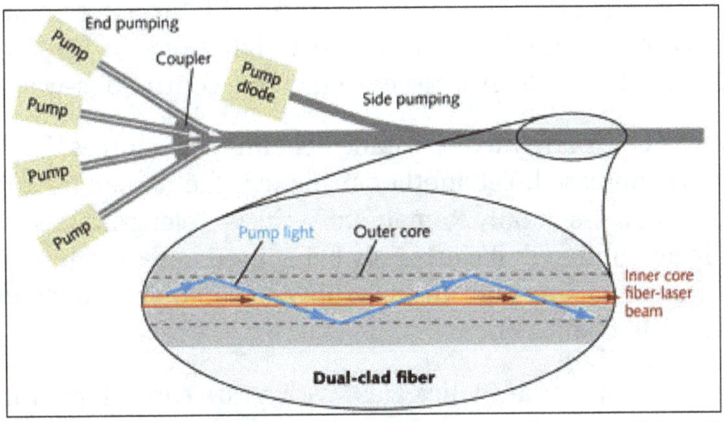

A fiber laser can be end-pumped with one or many lasers, or side-pumped (usually with many lasers) by side-coupling pump light into the outer core.

A fiber laser can be end- or side-pumped. In end-pumping, the light from one or many pump lasers is fired into the end of the fiber. In side-pumping, pump light is coupled into the side of the fiber; actually, it is fed into a coupler that couples it into the outer core. This is different from side-pumping a laser rod, where the light comes in orthogonally to the axis.

Many design considerations go into making this work. Considerable attention is spent on coupling the pump light into the core, matching it to the optical absorption, and coupling that pump light into the inner core to produce a population inversion that will result in stimulated emission in the inner core. The laser core can have various degrees of gain, depending on the doping in the fiber as well as on the fiber length. These are factors that the design engineer would adjust to get the performance that is needed.

Power limitations can arise, particularly from working within a single mode fiber. Such a fiber core has a very small cross-sectional area, and as a result, very high-intensity light going through it. Nonlinear Brillouin scattering becomes increasingly important at these high intensities, and can limit output at multi-kilowatt levels. If the output is high enough, the fiber end can be optically damaged.

Features of Fiber Lasers

Using a fiber as a laser medium gives a long interaction length, which works well for diode-pumping. This geometry results in high photon conversion efficiency, as well as a rugged and compact design. When fiber components are spliced together, there are no discrete optics to adjust or to get out of alignment.

The fiber-based laser design is highly adaptable. It can be adapted to do anything from welding heavy sheets of metal to producing femtosecond pulses. Many variations exist on the fiber-laser theme, as well as some configurations that are not, strictly speaking, fiber lasers. Fiber amplifiers provide single-pass amplification; they're used in telecommunications because they can amplify many wavelengths simultaneously. Fiber amplification is also used in the master-oscillator power-amplifier (MOPA) configuration, where the intent is to generate a higher output from a fiber laser. In some circumstances, an amplifier is used even with a continuous-wave (CW) laser.

Another example is fiber-amplified spontaneous-emission sources, in which the stimulated emission is suppressed. Yet another example is the Raman fiber laser, which relies on Raman gain that essentially Raman-shifts the wavelength. This is an application that's not in wide use, although it's certainly being used in research. In fact, some new research is going on into using fluoride glass fibers for Raman lasing and amplification rather than the standard silica fibers.

However, the fiber host is usually silica glass with a rare earth dopant in the core. The primary dopants are ytterbium and erbium. Ytterbium has center wavelengths ranging from about 1030 to 1080 nm and can emit in a broader range of wavelengths if pushed. Using pump diodes emitting in the 940 nm range can make the photon deficit very small. Ytterbium has none of the self-quenching effects that occur in neodymium at high densities, which is why neodymium is used in bulk lasers and ytterbium is used in fiber lasers (they both provide roughly the same wavelength).

Erbium fiber lasers emit at 1530 to 1620 nm, which is an eye-safe wavelength range. This can be frequency-doubled to generate light at 780 nm—a wavelength that's not available from fiber lasers in other ways. And finally, ytterbium can be added to erbium so that the ytterbium absorbs pump light and transfers that energy to erbium. Thulium is another dopant that emits even deeper into the near-infrared (NIR; 1750 to 2100 nm), and is thus another eye-safe material.

High Efficiency

Fiber lasers are quasi-three-level systems. A pump photon excites a transition from a ground state to an upper level; the laser transition is a drop from the lowest part of the upper level down into some of the split ground states. This is very efficient: For example, ytterbium with a pump photon at 940 nm produces an emitted photon at 1030 nm—a quantum defect (lost energy) of only about 9%.

Photon conversion efficiencies

Active element	Output	Pump bands	Photon conversion
Neodymium (solid-state)	1064 to 1088 nm	808 nm	76%
Ytterbium	1030 to 1100 nm	910, 940, 975 nm	Over 90% at 940/1030
Erbium	1550 nm band	980 and 1480 nm	95% or 63%
Yb-Er	Erbium	Ytterbium	
Thulium	1750 to 2100 nm	793 nm	2 out/1 in

In contrast, neodymium pumped on its standard 808 nm pump line has a quantum defect of about 24%. So ytterbium has an inherently higher efficiency, although not all that efficiency can be realized because some photons are lost. Ytterbium can be pumped in a number of bands. Erbium can be pumped at either 1480 or 980 nm; the latter is not as efficient from a photon defect point of view, but is useful even so because better pump sources are available at 980 nm.

Overall fiber-laser efficiency is the result of a two-stage process. First is the efficiency of the pump diode. Semiconductor lasers are very efficient, with on the order of 50% electrical-to-optical efficiency. Laboratory results are even better, with 70% or even more of the electrical pump energy being converted into light. When this output is matched carefully to the fiber laser's absorption line, the result is the pump efficiency.

The second is the optical-to-optical conversion efficiency. With a small photon defect, high excitation and extraction efficiency can be achieved, producing optical-to-optical conversion efficiency on the order of 60% to 70%. The result is wall-plug efficiency in the 25% to 35% range.

Configurations for Many Purposes

Continuous-wave fiber lasers can be either single- or multimode (in terms of transverse modes). A single mode produces a high-quality beam for materials working or sending a beam through the atmosphere, while multimode industrial lasers can generate higher raw power. If an application does not require the extremely high intensities resulting from single mode operation, the higher total power from multimode operation is often an advantage—for example, for some kinds of cutting and welding, and particularly for heat-treating, where a large area is illuminated.

Long-pulse fiber lasers are essentially quasi-CW lasers, typically producing millisecond-type pulses. Typically they have a 10% duty cycle (resulting from the pump diode modulation). This results in higher peak powers than in CW operation—typically on the order of ten times higher. This can be an advantage for some kinds of materials working such as pulse drilling. The repetition rate can range up to 500 Hz, depending on the pulse duration.

Q-switching is possible in fiber lasers, with the principle being the same as for bulk Q-switched lasers. Typical pulse lengths range from low nanosecond up to the microsecond range; the longer the fiber, the more time is needed to Q-switch the output, producing a longer pulse.

Fiber properties impose some limitations on Q-switching. Nonlinearities are more severe in a fiber laser due to the core's small cross-sectional area, so the peak power has to be somewhat limited. One can either use bulk Q-switches, giving higher performance, or a fiber Q-switch, which is spliced to the ends of the active part of the fiber laser.

The Q-switched pulses can be amplified in fiber or in bulk. An example of the latter is found at the National Ignition Facility (NIF; Livermore, CA), where a fiber laser is the master oscillator for the 192 beams of the NIF laser: Small pulses from the fiber laser are amplified up to mega-joule size in large slabs of doped glass.

In mode-locked fiber lasers, the repetition rate depends on the length of the gain material, as in any kind of mode-locking scheme, while pulse duration depends on the gain bandwidth. The shortest achievable oscillator pulses are in the 50 fs range, with more typical durations in the 100 fs range. Shorter pulses can be generated in oscillator-amplifier systems with external chirped-pulse amplification and subsequent pulse compression.

An important difference exists between erbium- and ytterbium-doped fibers, resulting from the fact that the two are operating in different dispersion modes. Erbium-doped fibers emit at 1550 nm, which is in the anomalous-dispersion region; this allows the production of solitons. Ytterbium-doped fibers are in the positive or normal dispersion realm; as a result, they generate strongly chirped pulses. As a result, a chirped fiber Bragg grating may be needed to dechirp the pulses to compress the pulse length.

There are a number of ways to modify fiber-laser pulses, particularly for things like picosecond ultrafast research. Photonic-crystal fibers can be made with extremely small cores to produce strong nonlinear effects for applications such as super continuum generation. In contrast, photonic crystals also can be made with very large single mode cores to avoid nonlinear effects at high power.

Bendable large-core photonic-crystal fibers are being created for high-power applications; one of the tricks being looked at is intentionally bending such a fiber enough that any undesired higher-order modes will go away, leaving only the fundamental transverse mode. Nonlinearities allow harmonics to be generated; sum and difference frequency mixing can create higher frequencies and shorter wavelengths. Nonlinear effects can also produce pulse compression, leading the way to the production of frequency combs.

In a super continuum source, very short pulses produce a broad continuous spectrum via self-phase modulation. In one example, initial 6 ps pulses at 1050 nm (from a ytterbium fiber laser) produces a spectrum ranging from the ultraviolet to beyond 1600 nm, with some irregularity across the spectrum. Another super continuum source, this one working in the IR, is pumped with an erbium fiber laser at 1550 nm. In this case, the spectrum of the super continuum source varies with pulse width. This design achieves a range of more than an octave, ranging out past 2200 nm.

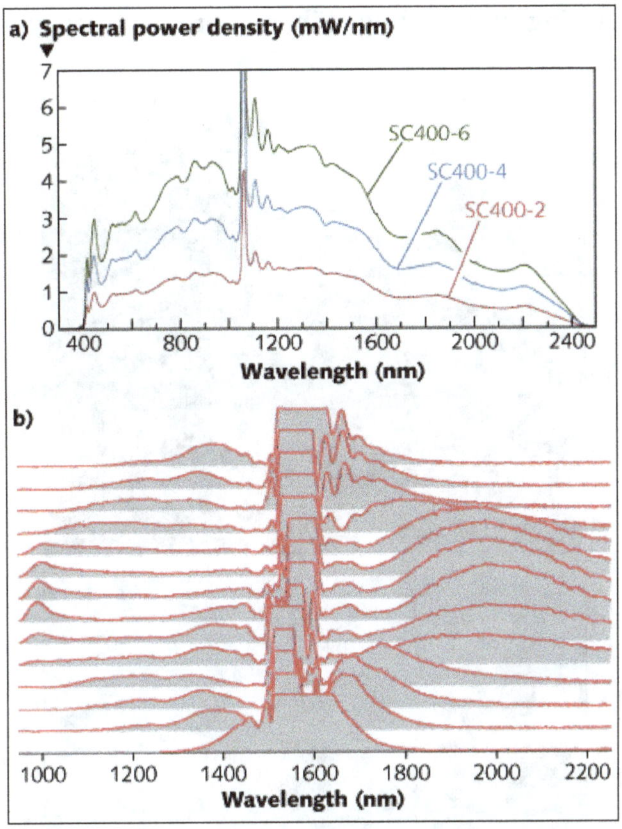

A fiber super continuum produces a broadband spectral output from narrowband pulses via nonlinear optical effects. Super continuum outputs in the watt range are achievable. The spectral output of an IR super continuum fiber can be changed by varying the pulse-width.

High-Power Fiber Lasers

The industrial market is now the largest market for fiber lasers; much of the action right now is at the kilowatt-class power level. Particularly interesting is their use in automotive work. The automotive industry is moving to high-strength steel to produce cars that meet durability requirements but are relatively light for better fuel economy; the problem is how to cut the high-strength steel. And that's where they turn to fiber lasers. It's very difficult, for example, for conventional machine tools to punch holes in this kind of steel; however, fiber lasers (and other types of lasers as well) can easily cut these holes.

Fiber lasers have some advantages over other lasers for materials processing. For example, the near-IR wavelengths of fiber lasers are absorbed well by metals. The beam can also be delivered by fiber, which allows a robot to easily move the beam focus around for cutting and drilling. Fiber lasers can satisfy extreme power requirements.

The US Navy's Laser Weapon System (LaWS) contains six individual fiber lasers with their beams incoherently combined into a single 33 kW output.

The highest single-mode power available from a fiber laser is 10 kW, from IPG Photonics. In the system, a master oscillator produces a kilowatt of optical power that is fed into an amplifier stage pumped at 1018 nm with light from other fiber lasers. The entire laser system is about the size of two refrigerators.

The highest multimode power reached is 50 kW, also by IPG Photonics. The system relies on incoherent beam combination, so it's not a super high-quality beam (beam parameter product of 10, M^2 of 33). This laser has been shipped around the country and has operated at 50 kilowatts in five states. This is the kind of durability that makes fiber lasers attractive for industry.

Other applications exist for fiber lasers in high-power cutting and welding—for example, replacing resistance welding for high-speed sheet steel, solving the problem of material distortion caused by resistance welding. Power and other feedback controls allow fiber lasers to cut a very precise curve, especially going around corners.

Drilling Concrete

A 4 kW multimode fiber laser has been used to cut and drill concrete. Why would one want to cut concrete with a laser? When engineers are attempting to earthquake-proof existing buildings, they want to be very careful with the concrete. However, they need to put things like rebar in it to bolster its strength so it won't just crumble if an earthquake hits. Conventional percussion drilling can crack and weaken concrete, but fiber lasers cut it without fracturing.

Q-switched fiber lasers are used, for example, in pulsed materials working, such as laser marking or working semiconductor electronics. They are also used for lidar; a module green module about the size of one's hand, contains an eye-safe erbium fiber laser with a 4 kW peak power, a 50 kHz repetition rate, and a 5-to-15-ns pulse duration.

There is a lot of interest in smaller fiber lasers for micro-and nano-scale machining. For surface ablation, if the pulse duration is made shorter than about 35ps, then there is no material splatter, just ablation, eliminating the formation of kerfs and other unwanted artifacts on the metal being cut. Pulses down in the femtosecond regime and produce nonlinear effects that are insensitive to wavelength and don't heat the surrounding area, allowing material work without damaging or weakening the surrounding area. In addition, holes can be cut with high aspect ratios—for example, rapidly (within a few milliseconds) drilling small holes through 1-mm-thick stainless steel using 800-fs pulses at a 1 MHz repetition rate.

One can also do surface machining of transparent materials—for example, the human eye. To cut flaps for LASIK surgery, femtosecond pulses are tightly focused with a high-numerical-aperture lens onto a spot below the eye's surface, causing no damage at the surface, but breakdown of the eye material at a controlled depth. The surface of the cornea, the smooth surface of which is important for vision, escapes unharmed. The flap, separated from underneath, can then be pulled up for ablative excimer-laser lens shaping. Other medical applications for picosecond and femtosecond fiber lasers include shallow-penetration surgery in dermatology, and use in certain kinds of optical coherence tomography (OCT).

Scientific applications of femtosecond fiber lasers include laser-induced breakdown spectroscopy, time-resolved fluorescence spectroscopy, and general materials research. And fiber lasers are used to produce femtosecond frequency combs, valuable in metrology as well as general research. One realistic near-term application for easily generated frequency combs will be atomic clocks for future-generation GPS satellites, resulting in much finer position resolution on the ground—a boon not just for knowing exactly where you are on the street, but also for automated equipment such as robot tractors that need to know they're tilling between, not on, rows of corn based on GPS measurements.

A single-frequency fiber laser is available with a line-width of less than 1 kHz. This is an impressively small device with an optical output of 10 mw to 1 W in the erbium fiber band. It is useful in communications, metrology (for example, in fiber gyroscopes), and spectroscopy.

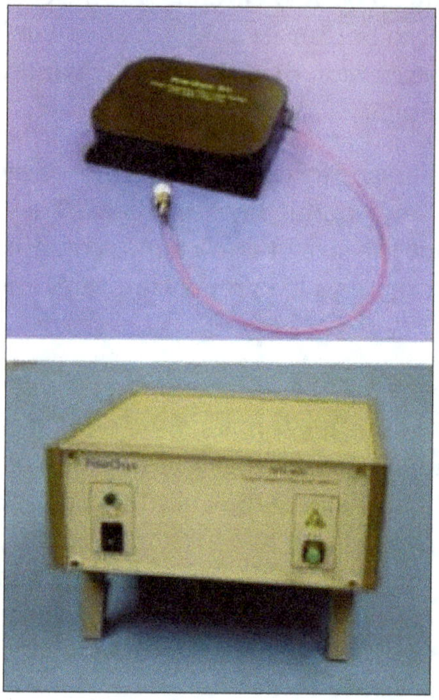

A single-frequency fiber laser has a linewidth of less than 1 kHz.

References

- Lasers: explainthatstuff.com, Retrieved 19, July 2020
- Laser-beam-characterization: rp-photonics.com, Retrieved 13, February 2020
- Principles-laser-diffraction: atascientific.com.au, Retrieved 06, August 2020
- Introduction-to-solidstate-lasers-16555507: laserfocusworld.com, Retrieved 21, January 2020
- Diode-Pumped-Lasers-Performance-Reliability-25160: photonics.com, Retrieved 12, April 2020
- Hybrid-Silicon-Lasers-252599749: researchgate.net, Retrieved 04, June 2020
- Fiber-lasers-fiber-lasers-the-state-of-the-art, spectroscopy, test-measurement: laserfocusworld.com, Retrieved 29, May 2020

2

Basic Theory of Laser

The working of laser depends on various theories such as interaction of light with matter, absorption and emission of light, Plank's quantum theory of light, Einstein coefficient relation, laser rate equation, population inversion and optical pumping. The diverse aspects and applications of these theories in relation to lasers have been thoroughly discussed in this chapter.

Interaction of Light with Matter

The interaction of light with matter can take many forms:

Absorption and Emission of Photons

Photons are annihilated or created.

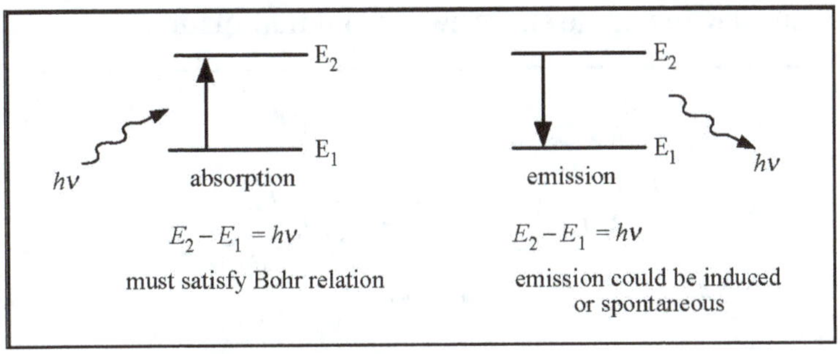

Most absorbing materials satisfy the Beer-Lambert Law.

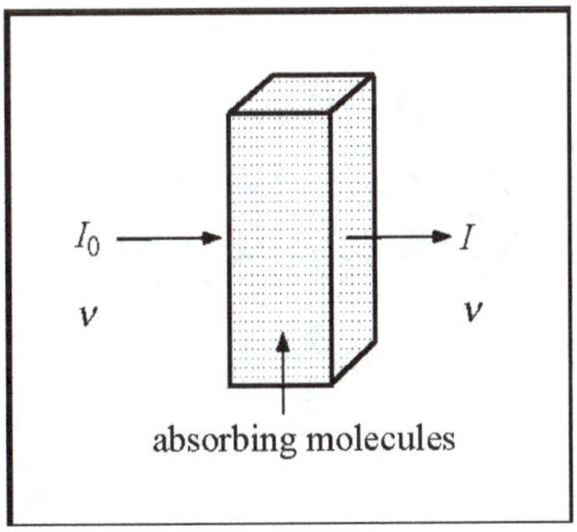

If,

I_o ≡ Incident light intensity at frequency ν.

I ≡ Intensity of transmitted light at same frequency.

C ≡ Concentration of absorbers.

ε ≡ Molar extinction coefficient.

l ≡ Length of light path through the absorbers.

Then:

$$\log_{10}\frac{I_0}{I} = \varepsilon l C$$

Quantity $\log_{10}\frac{I_0}{I} = A$ is called the absorbance or sometimes optical density (OD).

Light Scattering

An isolated atom scatters light because the electric field of the incident light wave forces the electrons in the atom to oscillate back and forth about their equilibrium position. By the laws of electromagnetism, when a charge changes its velocity, it emits radiation. Light is emitted uniformly in all directions in the plane ⊥ to oscillation, but decreases in amplitude as the viewing angle shifts away from that plane.

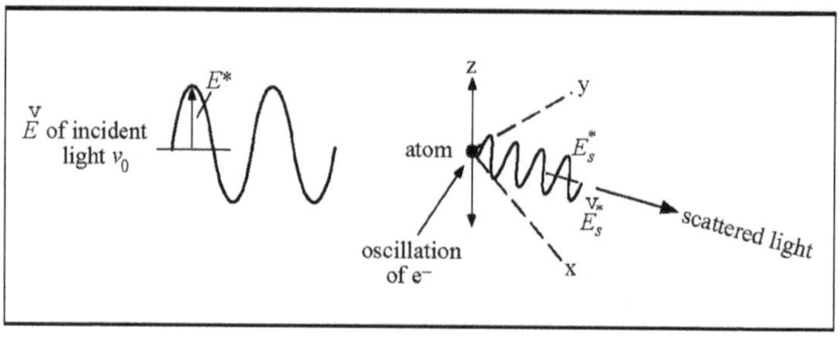

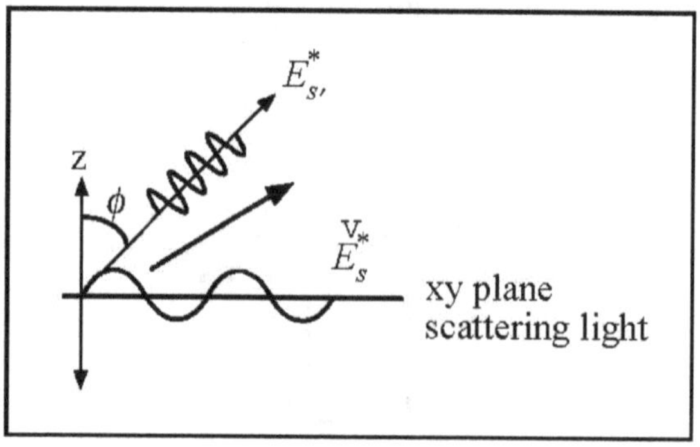

$$E_s^* = \frac{|e| a \sin\phi}{c^2 r}$$

Where,

$|-e|$ = Electron charge.

a = Acceleration of electron.

c = Velocity of light.

r = Distance from oscillator.

φ = Angle between axis of oscillation and plane containing direction of propagation of scattered light.

$$\text{Intensity of scattered light} \propto \left|E_2^*\right|^2$$

Scattered light is not necessarily at the same frequency as incident light.

- Rayleigh Scattering: Scattered light is essentially at the same frequency as incident light. Frequency shift due to:
    - Molecular motion → Doppler Shift.
    - Change in internal energy states of scattering molecule.

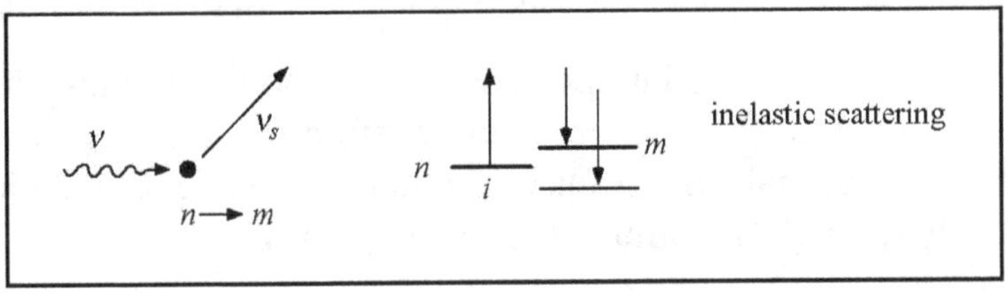

Inelastic scattering is important for gases, e.g., rotational Raman. In liquids, random molecular motions broaden Rayleigh scattering.

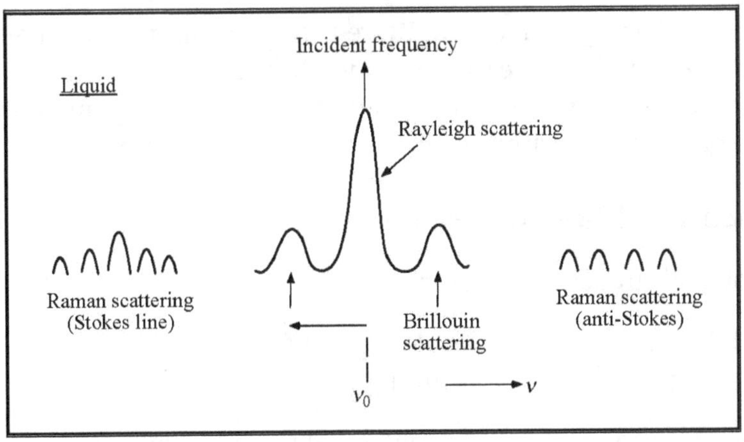

- Brillouin Scattering: In addition, in liquids, two special scattered lines, the

Brillouin lines, appear because of coherent molecular motion in sound waves that propagate through the fluid.

- Raman Scattering.

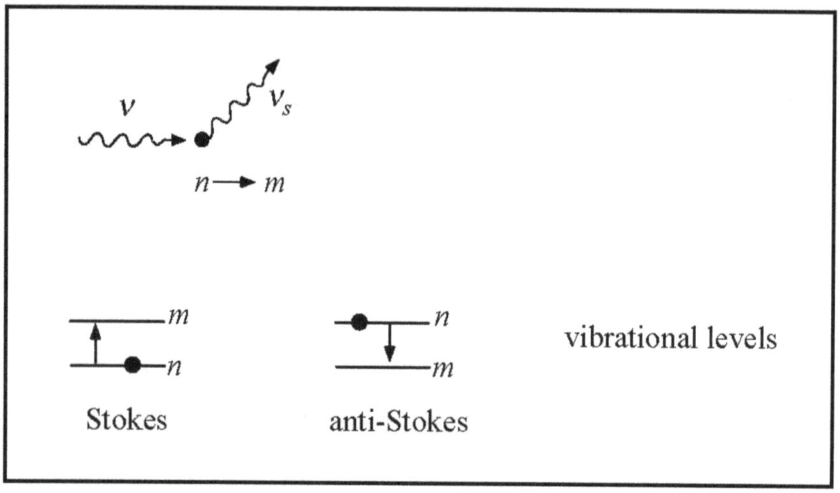

Phenomena Related to Light Scattering

- Reflection: Light scattered in the opposite direction of incident light.
- Refraction: Light scattered in the forward direction combines with the incident beam to give rise to the phenomenon of refraction. The physical effect of this combination is to make the transmitted light appear as though it has travelled more slowly through the sample than through a vacuum.

$$\text{index of refraction} = n = \frac{\text{velocity of light in vacuum}}{\text{velocity of light in substance}}$$

- Diffraction: Superposition of scattered waves from individual atoms or molecules in the sample. If the sample is highly ordered, diffraction pattern periodicity in the distribution of atoms and molecules in the sample can be used to deduce or infer the relative positions of atoms in a sample.

Fluorescence and Phosphorescence

Optical rotation and circular dichroism:

- Optical Rotation: Chiral molecules can rotate the plane of polarization of plane polarized light as light passes through the sample.
- Circular Dichroism: Chiral molecules will absorb right and left circularly polarized light differentially.

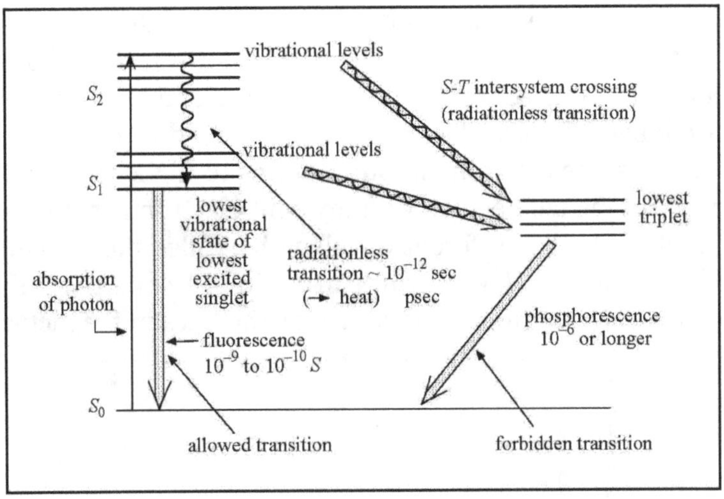

Absorption and Emissions of Light

The electrons surrounding the atomic nucleus are arranged in a series of levels of increasing energy. Each element has a unique number of electrons in a unique configuration therefore each element has its own distinct set of energy levels. This arrangement of energy levels serves as the atom's unique fingerprint.

In the early 1900s, scientists found that a liquid or solid heated to high temperatures would give off a broad range of colours of light. However, a gas heated to similar temperatures would emit light only at certain specific wavelengths (colours). The reason for this observation was not understood at the time. Scientists studied this effect using a discharge tube.

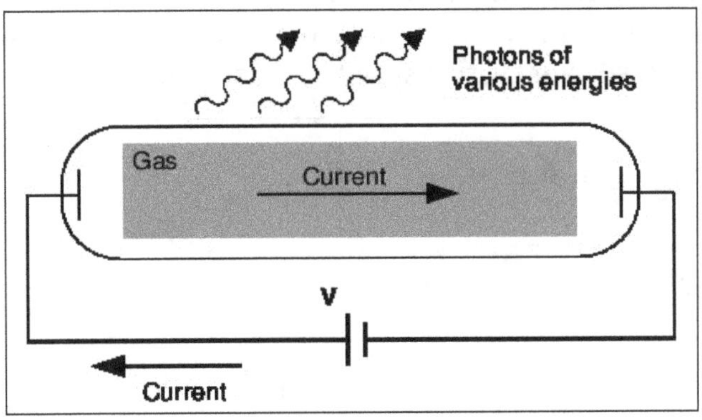

The tube is filled with a gas. When a high enough voltage is applied across the tube, the gas ionises and acts like a conductor, allowing a current to flow through the circuit. The current excites the atoms of the ionised gas. When the atoms fall back to their ground state, they emit photons to carry off the excess energy.

A discharge tube is a gas-filled, glass tube with a metal plate at both ends. If a large enough voltage difference is applied between the two metal plates, the gas atoms inside the tube will absorb enough energy to make some of their electrons come off, i.e. the gas atoms are ionised. These electrons start moving through the gas and create a current, which raises some electrons in other atoms to higher energy levels. Then as the electrons in the atoms fall back down, they emit electromagnetic radiation (light). The amount of light emitted at different wavelengths, called the emission spectrum, is shown for a discharge tube filled with hydrogen gas in Figure below. Only certain wavelengths (i.e. colours) of light are seen, as shown by the lines in the picture.

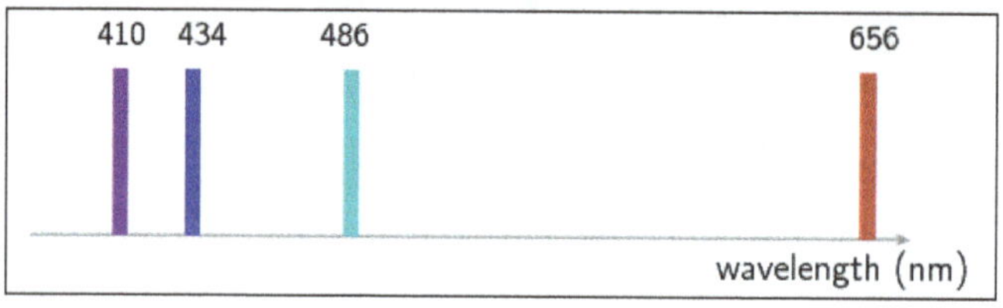

Diagram of the emission spectrum of hydrogen in the visible spectrum. Four lines are visible, and are labelled with their wavelengths. The three lines in the 400–500 nm range are in the blue part of the spectrum, while the higher line (656 nm) is in the red/orange part.

Eventually, scientists realised that these lines come from photons of a specific energy, emitted by electrons making transitions between specific energy levels of the atom. Figure below shows an example of this happening. When an electron in an atom falls from a higher energy level to a lower energy level, it emits a photon to carry off the extra energy. This photon's energy is equal to the energy difference between the two energy levels (ΔE).

$$\Delta E\text{electron} = E_f - E_i$$

The frequency of a photon is related to its energy through the equation *E=hf*. Since a specific photon frequency (or wavelength) gives us a specific colour, we can see how each coloured line is associated with a specific transition.

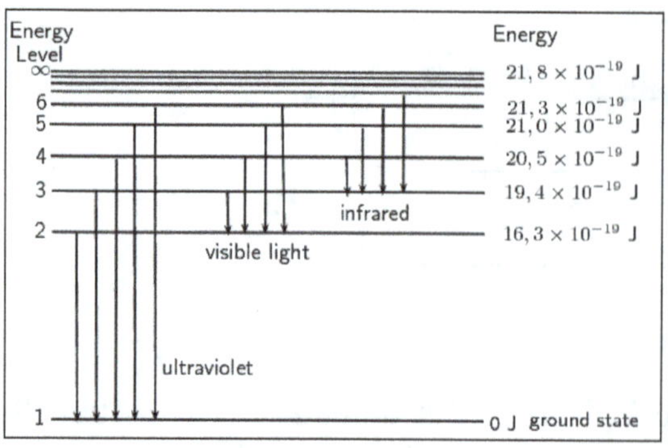

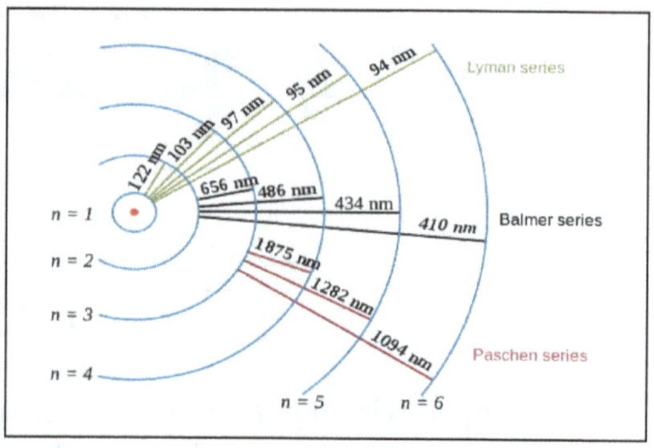

In the first diagram are shown some of the electron energy levels for the hydrogen atom. The arrows show the electron transitions from higher energy levels to lower energy levels. The energies of the emitted photons are the same as the energy difference between two energy levels. You can think of absorption as the opposite process. The arrows would point upwards and the electrons would jump up to higher levels when they absorb a photon of the right energy. The second representation shows the wavelengths of the light that is emitted for the the various transitions. The transistions are grouped into a series based on the lowest level involved in the transition.

Absorption Spectra

Atoms do not only emit photons; they also absorb photons. If a photon hits an atom and the energy of the photon is the same as the gap between two electron energy levels in the atom, then the electron in the lower energy level can absorb the photon and jump up to the higher energy level. If the photon energy does not correspond to the difference between two energy levels then the photon will not be absorbed (it can still be scattered).

Using this effect, if we have a source of photons of various energies we can obtain the absorption spectra for different materials. To get an absorption spectrum, just shine white light on a sample of the material that you are interested in. White light is made up of all the different wavelengths of visible light put together. In the absorption spectrum there will be gaps. The gaps correspond to energies (wavelengths) for which there is a corresponding difference in energy levels for the particular element.

The absorbed photons show up as black lines because the photons of these wavelengths have been absorbed and do not show up. Because of this, the absorption spectrum is the exact inverse of the emission spectrum. Look at the two figures below. In Figure below you can see the line emission spectrum of hydrogen. Figure below shows the absorption spectrum. It is the exact opposite of the emission spectrum. Both emission and absorption techniques can be used to get the same information about the energy levels of an atom.

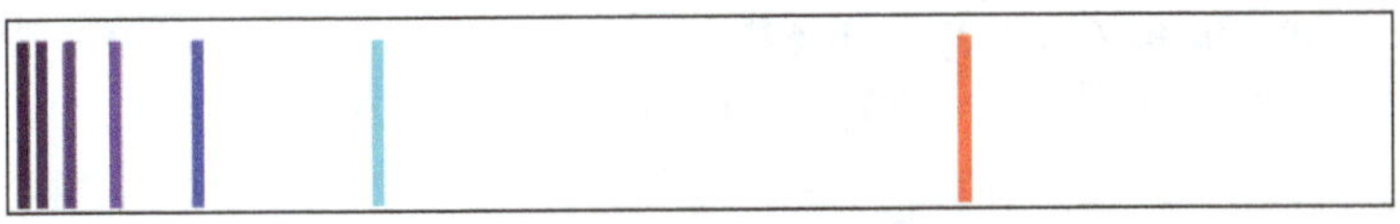

Emission spectrum of Hydrogen.

Absorption spectrum of Hydrogen.

The dark lines correspond to the frequencies of light that have been absorbed by the gas. As the photons of light are absorbed by electrons, the electrons move into higher energy levels. This is the opposite process of emission.

The dark lines, absorption lines, correspond to the frequencies of the emission spectrum of the same element. The amount of energy absorbed by the electron to move into a higher level is the same as the amount of energy released when returning to the original energy level.

The energy of the photon does not correspond to the energy of an energy level, it corresponds to the difference in energy between two energy levels.

Applications of Emission and Absorption Spectra

The study of spectra from stars and galaxies in astronomy is called spectroscopy. Spectroscopy is a tool widely used in astronomy to learn different things about astronomical objects.

Identifying Elements in Astronomical Objects using their Spectra

Measuring the spectrum of light from a star can tell astronomers what the star is made of. Since each element emits or absorbs light only at particular wavelengths, astronomers can identify what elements are in the stars from the lines in their spectra. From studying the spectra of many stars we know that there are many different types of stars which contain different elements and in different amounts.

Determining Velocities of Galaxies using Spectroscopy

The same thing happens to electromagnetic radiation (light). If the object emitting the light is moving towards us, then the wavelength of the light appears shorter (called blue-shifted). If the object is moving away from us, then the wavelength of its light appears stretched out (called red-shifted).

The Doppler effect affects the spectra of objects in space depending on their motion relative to us on the earth. For example, the light from a distant galaxy that is moving away from us at some velocity will appear red-shifted. This means that the emission and absorption lines in the galaxy's spectrum will be shifted to a longer wavelength (lower frequency). Knowing where each line in the spectrum would normally be if the galaxy was not moving and comparing it to the red-shifted position, allows astronomers to precisely measure the velocity of the galaxy relative to the Earth.

Global Warming and Greenhouse Gases

The sun emits radiation (light) over a range of wavelengths that are mainly in the visible part of the spectrum. Radiation at these wavelengths passes through the gases of the atmosphere to warm the land and the oceans below. The warm earth then radiates this heat at longer infrared wavelengths. Carbon dioxide (one of the main greenhouse gases) in the atmosphere has energy levels that correspond to the infrared wavelengths that allow it to absorb the infrared radiation. It then also emits at infrared wavelengths in all directions. This effect stops a large amount of the infrared radiation from getting out of the atmosphere, which causes the atmosphere and the earth to heat up. More radiation is coming in than is getting back out.

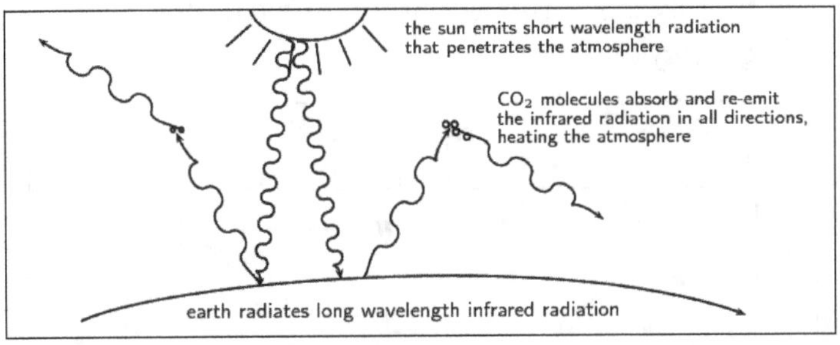

So increasing the amount of greenhouse gases in the atmosphere increases the amount of trapped infrared radiation and therefore the overall temperature of the earth. The earth is a very sensitive and complicated system upon which life depends and changing the delicate balances of temperature and atmospheric gas content may have disastrous consequences if we are not careful.

Planck's Quantum Theory of Light

In the late 18th century, great progress in physics had been made. Classical Newtonian physics at the time was widely accepted in the scientific community for its ability to accurately explain and predict many phenomena. However, by the early 20th century, physicists discovered that the laws of classical mechanics are not applicable at the atomic scale, and experiments such as the photoelectric effect completely contradicted the laws of classical physics. As a result of these observations, physicists articulated a set of theories now known as quantum mechanics. In some ways, quantum mechanics completely changed the way physicists viewed the universe, and it also marked the end of the idea of a clockwork universe (the idea that universe was predictable).

Electromagnetic Radiation

Electromagnetic (EM) radiation is a form of energy with both wave-like and particle-like properties; visible light being a well-known example. From the wave perspective, all forms of EM radiation may be described in terms of their wavelength and frequency. Wavelength is the distance from one wave peak to the next, which can be measured in meters. Frequency is the number of waves that pass by a given point each second. While the wavelength and frequency of EM radiation may vary, its speed in a vacuum remains constant at 3.0×10^8 m/sec, the speed of light. The wavelength or frequency of any specific occurrence of EM radiation determine its position on the electromagnetic spectrum and can be calculated from the following equation:

[latex]c = \lambda \nu[/latex]

where c is the constant 3.0×10^8 m/sec (the speed of light in a vacuum), [latex]\lambda[/latex] = wavelength in meters, and [latex]\nu[/latex]=frequency in hertz (1/s). It is important to note that by using this equation, one can determine the wavelength of light from a given frequency and vice versa.

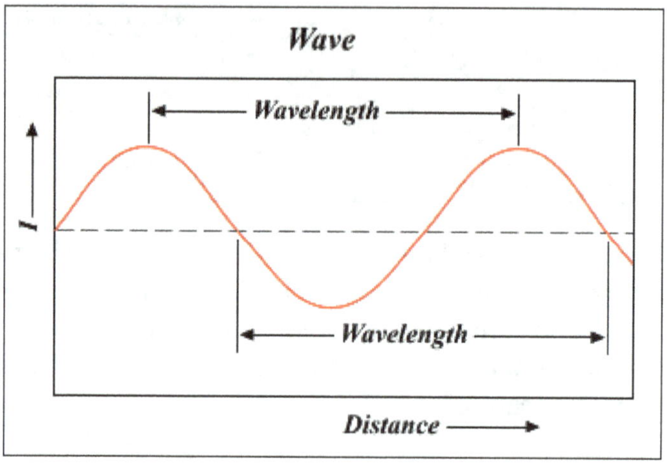

Wavelength of EM radiation: The distance used to determine the wavelength is shown. Light has many properties associated with its wave nature, and the wavelength in part determines these properties.

The Discovery of the Quantum

The wave model cannot account for something known as the photoelectric effect. This effect is observed when light focused on certain metals emits electrons. For each metal, there is a minimum threshold frequency of EM radiation at which the effect will occur. Replacement of light with twice the intensity and half the frequency will not produce the same outcome, contrary to what would be expected if light acted strictly as a wave. In that case, the effect of light would be cumulative—the light should add up, little by little, until it caused electrons to be emitted. Instead, there is a clear-cut minimum frequency of light that triggers electron ejection. The implication was that frequency is directly proportional to energy, with the higher light frequencies having more energy. This observation led to the discovery of the minimum amount of energy that could be gained or lost by an atom. Max Planck named this minimum amount the "quantum," plural "quanta," meaning "how much." One photon of light carries exactly one quantum of energy.

Planck is considered the father of the Quantum Theory. According to Planck: E=h[latex]\nu[/latex], where h is Planck's constant ($6.62606957(29) \times 10^{-34}$ J s), ν is the frequency, and E is energy of an electromagnetic wave. Planck (cautiously) insisted that this was simply an aspect of the processes of absorption and emission of radiation and had nothing to do with the physical reality of the radiation itself. However, in 1905, Albert Einstein reinterpreted Planck's quantum hypothesis and used it to explain the photoelectric effect, in which shining light on certain materials can eject electrons from the material.

When an electric current is passed through a gas, some of the electrons in the gas molecules move from their ground energy state to an excited state that is further away from their nuclei. When the electrons return to the ground state, they emit energy of various wavelengths. A prism can be used to separate the wavelengths, making them easy to identify. If light acted only as a wave, then there should be a continuous rainbow created by the prism. Instead, there are discrete lines created by different wavelengths. This is because electrons release specific wavelengths of light when moving from an excited state to the ground state.

Emission spectrum of nitrogen gas: Each wavelength of light emitted (each colored line) corresponds to a transition of an electron from one energy level to another, releasing a quantum of light with defined energy (color).

Planck's Radiation Law

Planck's radiation law is a law of physics which explains the spectral energy distribution of the heat radiation emitted from a so-called blackbody at any temperature. Discovered by Max Planck, this law laid the foundation for the advent of the quantum theory because it was the first physical law to postulate that electromagnetic energy exists in discrete bundles, or quanta.

According to Maxwell's theory, an electromagnetic wave is a coupled self-sustaining oscillation of electric and magnetic fields that propagates though a vacuum at the speed of light, $c=3\times10^{8}\ ms^{-1}$. The electric component of the wave can be written:

$$\mathbf{E}=\mathbf{E}_0 \exp\left[i(k.r-\omega t\right],$$

where $\mathbf{E}_0$ is a constant, **k** is the wave vector that determines the wavelength and direction of propagation of the wave, and ω is the angular frequency. The dispersion relation:

$$\omega=kc$$

It ensures that the wave propagates at the speed of light. Note that this dispersion relation is very similar to that of sound waves in solids. Electromagnetic waves always propagate in the direction perpendicular to the coupled electric and magnetic fields (i.e., electromagnetic waves are transverse waves). This means that $\mathbf{k.E}_0=0$. Thus, once k is specified, there are only two possible independent directions for the electric field. These correspond to the two independent polarizations of electromagnetic waves.

Consider an enclosure whose walls are maintained at fixed temperature T. What is the nature of the steady-state electromagnetic radiation inside the enclosure? Suppose that the enclosure is a parallelepiped with sides of lengths L_x, L_y, and L_z. Alternatively (and more conveniently), suppose that the radiation field inside the enclosure is periodic in the x-, y-, and z-directions, with periodicity lengths L_x, L_y, and L_z, respectively. As long as the smallest of these lengths, L, say, is much greater than the longest wavelength of interest in the problem, $\lambda=2\pi/k$, then these assumptions should not significantly affect the nature of the radiation inside the enclosure. We find, that the periodicity constraints ensure that there are only a discrete set of allowed wavevectors (i.e., a discrete set of allowed modes of oscillation of the electromagnetic field inside the enclosure). Let $p(\mathbf{k})d^3\mathbf{k}$ be the number of allowed modes per unit volume with wavevectors in the range k to k + dk. We know, by analogy, that:

$$p(k)d^3k=\frac{d^3\mathbf{k}}{(2\pi)^3}.$$

The number of modes per unit volume for which the magnitude of the wavevector lies in the k to k+ dk is just the density of modes, p(k), multiplied by the ``volume" in k -space of the spherical shell lying between radii k and k+dk. Thus,

$$p\ (k)dk=\frac{4\ k^2dk}{(\ )}=\frac{k^2}{}dk.$$

Finally, the number of modes per unit volume whose frequencies lies between ω and $\omega+d\omega$ is, by $\omega=kc$:

$$\sigma(\omega)d\omega=2p_k(k)\frac{dk}{d\omega}d\omega=2\frac{\omega^2}{2\pi^2c^3}d\omega.$$

Here, the additional factor 2 is to take account of the two independent polarizations of the electromagnetic field for a given wave vector, k.

Let us consider the situation classically. By analogy with sound waves, we can treat each allowable mode of oscillation of the electromagnetic field as an independent harmonic oscillator. According to the equipartition theorem, each mode possesses a mean energy kT in thermal equilibrium at temperature T. In fact, (1/2) KT resides with the oscillating electric field, and another (1/2)Kt with the oscillating magnetic field. Thus, the classical energy density of electromagnetic radiation (i.e., the energy per unit volume associated with modes whose frequencies lie in the range $\omega+d\omega$) is:

$$\bar{u}(\omega)d\omega=kT\sigma(\omega)=\frac{kT}{\pi^2c^3}\omega^2\,d\omega.$$

This result is known as the Rayleigh-Jeans radiation law, after Lord Rayleigh and James Jeans who first proposed it in the late nineteenth century.

According to Debye theory, the energy density of sound waves in a solid is analogous to the Rayleigh-Jeans law, with one very important difference. In Debye theory, there is a cut-off frequency (the Debye frequency) above which no modes exist. This cut-off comes about because of the discrete nature of solids (i.e., because solids are made up of atoms instead of being continuous). It is, of course, impossible to have sound waves whose wavelengths are much less than the inter-atomic spacing. On the other hand, electromagnetic waves propagate through a vacuum, which possesses no discrete structure. It follows that there is no cut-off frequency for electromagnetic waves, and so the Rayleigh-Jeans law holds for all frequencies. This immediately poses a severe problem. The total classical energy density of electromagnetic radiation is given by:

$$U\int_0^{\infty}\bar{u}(\omega)\,d\omega=\frac{kT}{\pi^2c^3}\int_0^{\infty}\omega^2 d\omega.$$

This is an integral that obviously does not converge. Thus, according to classical physics, the total energy density of electromagnetic radiation inside an enclosed cavity is infinite. This is clearly an absurd result, and was recognized as such in the latter half of the nineteenth century. In fact, this prediction is known as the ultra-violet catastrophe, because the Rayleigh-Jeans law usually starts to diverge badly from experimental observations (by over-estimating the amount of radiation) in the ultra-violet region of the spectrum.

So, how do we obtain a sensible answer? As usual, quantum mechanics comes to our rescue. According to quantum mechanics, each allowable mode of oscillation of the electromagnetic field corresponds to a photon state with energy and momentum:

$$\epsilon=\hbar\omega,$$

$$p=\hbar k,$$

respectively. Incidentally, it follows from Equation $\omega=kc$ that,

$$\epsilon=pc,$$

which implies that photons are massless particles that move at the speed of light. According to the Planck distribution, the mean number of photons occupying a photon state of ω is,

$$n_{(\omega)}=\hbar\omega n_{\omega}(\omega)=\frac{\hbar\omega}{e^{\beta\hbar\omega}-1}.$$

Hence, the mean energy of such a state is given by:

$$\bar{\epsilon}(\omega)=\hbar\omega n_{\omega}(\omega)=\frac{\hbar\omega}{e^{\beta\hbar\omega}-1}.$$

Note that low-frequency states (i.e. $\hbar\omega \ll kT$) behave classically: that is,

$$\bar{\epsilon} \simeq kT.$$

On the other hand, high-frequency states (i.e. $\hbar\omega \ll kT$) are completely ``frozen out": that is,

$$\bar{\epsilon} \ll kT.$$

The reason for this is simply that it is very difficult for a thermal fluctuation to create a photon with energy greatly in excess of kT, because kT is the characteristic energy associated with such fluctuations.

The true energy density of electromagnetic radiation inside an enclosed cavity is written:

$$\bar{u}(\omega)d\omega = \bar{\epsilon}(\omega)d\omega,$$

Giving,

$$\bar{u}(\omega)d\omega = \frac{\hbar}{\pi^2 c^3}\frac{\omega^3\, d\omega}{\exp(\beta\hbar\omega)-1}.$$

This is famous result is known as the Planck radiation law. The Planck law approximates to the classical Rayleigh-Jeans law for $\hbar\omega \ll kT$ peaks at about $\hbar\omega \simeq 3kT$ and falls off exponentially for $\hbar\omega \ll kT$. The exponential fall off at high frequencies ensures that the total energy density remains finite.

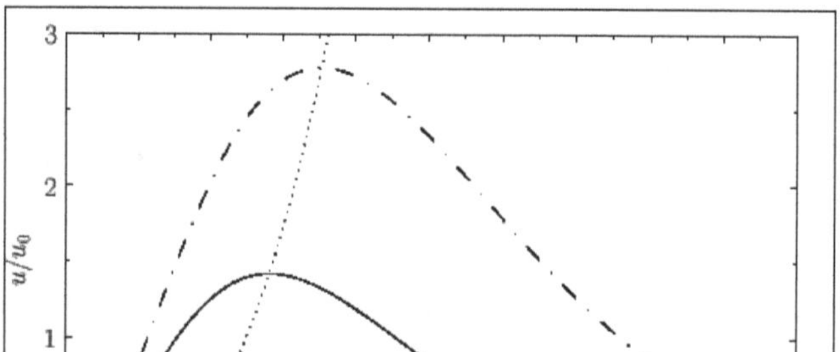

The Planck radiation law: Here, $\omega_0 = kT_0 / K$ and $u_0 = \hbar\omega_0^3 / (\pi^2 c^3)$, where T_0 is an arbitrary scale temperature. The dashed, solid, and dash-dotted curves show $\bar{u} / u_0$ for $T / T_0 = 0.75$, 1.0, and 1.25, respectively. The dotted curve shows the locus of the peak emission frequency.

Einstein Coefficient Relation

Einstein showed the interaction of radiation with matter with the help of three processes called stimulated absorption, spontaneous emission and stimulated emission. He showed in 1917 that for proper description of radiation with matter, the process of stimulated emission is essential. Let us first derive the Einstein coefficient relation on the basis of above theory.

Let N_1 be the number of atoms per unit volume in the ground state E_1 and these atoms exist in the radiation field of photons of energy E_2-E_1 =h v such that energy density of the field is E. Let R_1 be the rate of absorption of light by E_1 -> E_2 transitions by the process called stimulated absorption.

This rate of absorption R_1 is proportional to the number of atoms N_1 per unit volume in the ground state and proportional to the energy density E of radiations.

That is,

$$R_1 \infty N_1 E$$

Or,

$$R_1 = B_{12} N_1 E$$

Where B_{12} is known as the Einstein's coefficient of stimulated absorption and it represents the probability of absorption of radiation. Energy density e is defined as the incident energy on an atom as per unit volume in a state.

Now atoms in the higher energy level E_2 can fall to the ground state E_1 automatically after 10^{-8} sec by the process called spontaneous emission. The rate R_2 of spontaneous emission E_2-> E_1 is independent of energy density E of the radiation field. R_2 is proportional to number of atoms N_2 in the excited state E_2 thus,

$$R_2 \infty N_2,$$

$$R_2 = A_{21} N_2$$

Where A_{21} is known as Einstein's coefficient for spontaneous emission and it represents the probability of spontaneous emission.

Atoms can also fall back to the ground state E_1 under the influence of electromagnetic field of incident photon of energy E_2-E_1 =hv by the process called stimulated emission. Rate R_3 for stimulated emission E_2-> E_1 is proportional to energy density E of the radiation field and proportional to the number of atoms N_2 in the excited state, thus,

Where B_{21} is known as the Einstein coefficient for stimulated emission and it represents the probability of stimulated emission. In steady state (at thermal equilibrium), the two emission rates (spontaneous and stimulated) must balance the rate of absorption. Thus,

$$R_1=R_2+R_3$$

Using above equations, we get:

$$N_1B_{12}E=N_2A_{21}+N_2B_{21}E$$

Or

$$N_1B_{12}E-N_2B_{21}E=N_2A_{21}$$

Or

$$(N_1B_{12}-N_2B_{21})\,E=N_2A_{21}$$

Or

$$E=N_2A_{21}/N_1B_{12}-N_2B_{21}$$

$$=N_2A_{21}/N_2B_{21}[N_1B_{12}/N_2B_{21}-1]$$

by taking out common N_2B_{21} from the denominator:

$$E=A_{21}/B_{21}\{1/N_1/N_2(B_{12}/B_{21}-1)).$$

Einstein proved thermodynamically, that the probability of stimulated absorption is equal to the probability of stimulated emission. Thus,

$$B_{12}=B_{21}$$

Then $E=A_{21}/B_{21}\{1/N_1/N_2(B_{12}/B_{21}-1))$ becomes,

$$E=A_{21}/B_{21}(1/N_1/N_2-1)$$

From Boltzman's distribution law, the ratio of populations of two levels at temperature T is expressed as:

$$N_1/N_2=e^{(E2-E1)/KT}$$

$$N_1/N_2=e^{hv/KT}.$$

Where K is the Boltzman's constant and h is the Planck's constant. Substituting value of N1/N2in equation $E=A_{21}/B_{21}(1/N_1/N_2-1)$ we get,

$$E=A_{21}/B_{21}(1/e^{hv/KT}-1).$$

Now according to Planck's radiation law, the energy density of the black body radiation of frequency v at temperature T is given as,

$$E=8\pi hv^3/c^3(1/e^{hv/KT})$$

By comparing both the above equations, we get,

$$A_{21}/B_{21}=8\pi hv^3/c^3$$

This is the relation between Einstein's coefficients in laser. Significance of Einstein coefficient relation: This shows that the ratio of Einstein's coefficient of spontaneous emission to the Einstein's coefficient of stimulated absorption is proportional to cube of frequency v. It means that at thermal equilibrium, the probability of spontaneous emission increases rapidly with the energy difference between two states.

Einstein A and B Coefficients

In 1917, about 9 years before the development of the relevant quantum theory, Einstein postulated on thermodynamic grounds that the probability for spontaneous emission, A, was related to the probability of stimulated emission, B, by the relationship:

$$A / B=8\pi hv^3 / c^3.$$

From the development of the theory behind blackbody radiation, it was known that the equilibrium radiation energy density per unit volume per unit frequency was equal to:

$$p(v)=8\pi hv^3 / c^3.$$

Einstein argued that equilibrium would be possible, and the laws of thermodynamics obeyed, only if the ratio of the A and B coefficients had the value shown above. This ratio was calculated from quantum mechanics in the mid 1920's. In recognition of Einstein's insight, the coefficients have continued to be called the Einstein A and B coefficients.

While applicable in many situations, the A and B coefficients received particular attention in the period in which lasers were being developed. The nature of the coefficients is such that you cannot use the radiation in a cavity to elevate electrons preferentially into an upper state, producing the population inversion necessary for laser action. The particular ratio between the coefficients suggests that the presence of the light to "pump" electrons into upper states will have the same probability of stimulating an already elevated electron to make the downward transition, so that laser action cannot be achieved with any two-level system. The achievement of laser action was obtained by three-level systems like that in the helium-neon laser where the population of the upper neon level could be achieved by a non-radiative transfer from the helium gas pumping energy to the neon atoms.

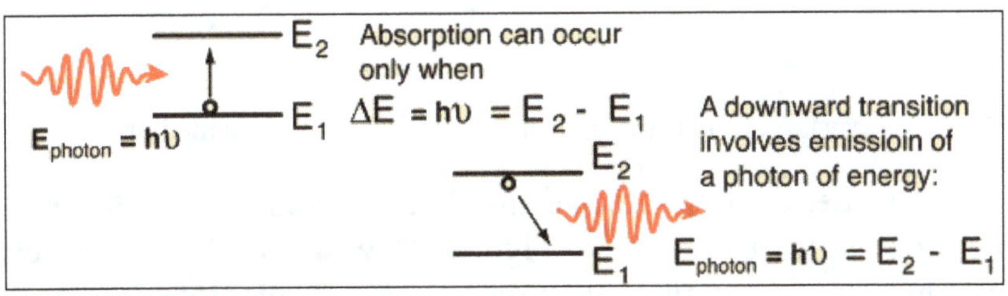

The implication of the Einstein A and B coefficients is that these two processes will occur at equal rates, so that no population inversion can be attained in a two-level. In the helium-neon laser, for example, this limitation is overcome by collisional transfer from the helium gas to the neon gas, achieving the necessary population inversion for laser action.

Laser Rate Equation

The dynamics of energy level populations in laser gain media (e.g. rare-earth-doped crystals or fibers) are often modeled using a system of rate equations. These are differential equations, describing the temporal evolution of level populations under the influence of optically induced and non-radiative transitions:

- Absorption processes, possibly including excited-state absorption.
- Spontaneous and stimulated emission.
- Multi-phonon transitions.
- Energy transfers which can lead to up conversion and quenching.

Rate equation models can be part of more comprehensive numerical models, which describe e.g. the spatial distribution of optical powers in fiber amplifiers or bulk lasers, or the dynamic behavior of Q-switched lasers. They can thus help in understanding quantitatively the operation of laser and amplifier devices, and allow one, e.g., to evaluate whether the performance of a device is close to the limitations set by the fundamental principle of operation.

Rate Equations for Erbium-Doped Gain Media

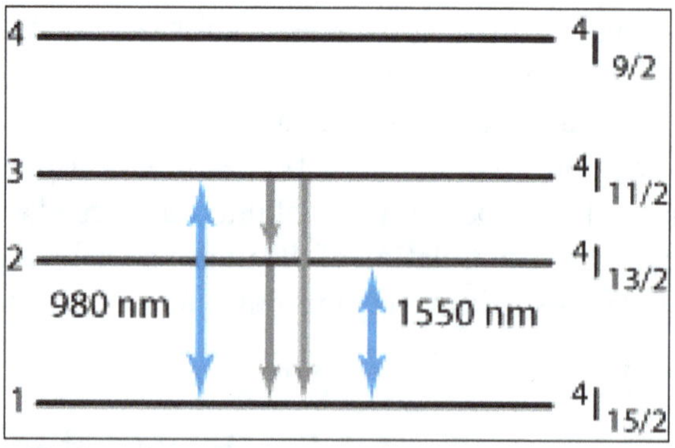

Energy levels and transitions in an erbium-doped gain medium.

As an example, consider the dynamics of an erbium-doped gain medium, such as used in, e.g., erbium-doped fiber amplifiers. Figure below shows the energy level diagram and the most important radiative and non-radiative transitions. Erbium amplifiers and lasers operating in the 1.5-µm spectral region exhibit a quasi-three-level scheme.

For simplicity, the energy levels (actually Stark level manifolds) of the erbium (Er^{3+}) ions are labeled on the left-hand side with an index which is e.g. 1 for the ground-state manifold $^4I_{15/2}$ and 2 for the upper-state manifold $^4I_{13/2}$. Optical absorption and stimulated emission transitions, as caused by a pump beam at 980 nm and a signal beam at 1550 nm, are indicated with blue arrows, and the gray arrows indicate spontaneous and non-radiative transitions. The lower transition (level 2 → 1) is mostly caused by spontaneous emission, whereas the upper transition (3 → 2) is dominated by multi-phonon transitions. Only transitions between the lowest three levels are considered, assuming that excited-state absorption to higher levels and up conversion processes are weak.

For that situation, the rate equation system obtained is the following:

$$\frac{\partial n_3}{\partial t} = -A_{32}n_3 - A_{31}n_3 + \frac{\sigma_{13}I_p}{h\nu_p}n_1 - \frac{\sigma_{31}I_s}{h\nu_p}n_3$$

$$\frac{\partial n_2}{\partial t} = A_{21}n_2 + A_{32}n_3 + \frac{\sigma_{12}I_s}{h\nu_s}n_1 - \frac{\sigma_{21}I_s}{h\nu_s} - \frac{\sigma_{21}I_s}{h\nu_s}n_2$$

$$\frac{\partial n_1}{\partial t} = +A_{21}n_2 + A_{31}n_3 - \frac{\sigma_{13}I_p}{h\nu_p}n_1 + \frac{\sigma_{31}I_p}{h\nu_p}n_3 - \frac{\sigma_{12}I_s}{h\nu_s}n_1 + \frac{\sigma_{21}I_s}{h\nu_s}n_2,$$

where n_j indicates the fractional level population of level j. This variable is e.g. 1 if all ions are in the corresponding level manifold. As only the mentioned three levels are involved, we have $n_1 + n_2 + n_3 = 1$. (The parameters n_j can also be interpreted as excitation densities with units of m^{-3}; only in rate equation systems containing nonlinear terms, e.g. for energy transfers, the form of the equations depends on that interpretation.) The parameters A_{jk} indicate spontaneous transition rates from level j to k, with units of s^{-1}. For example, A_{21} is the inverse upper-state lifetime. Furthermore, the equations contain absorption and stimulated emission rates, which are determined by transition cross sections σ_{jk} (the values of which are dependent on the wavelengths involved), optical intensities I_p and I_s at the pump and signal wavelength, and photon energies $h\nu$.

The multi-phonon transition 3 → 2 is usually strong if the gain medium (e.g. a silica fiber) has high phonon energy. Therefore, n_3 will usually be small (except for extremely high pump intensities), as ions pumped into level 3 will rapidly be transferred to level 2. Consequently, it is often valid to neglect the population in level 3, and also the other transitions starting from that level. The equation system then simplifies to:

$$\frac{\partial n_3}{\partial t} = -A_{21}n_2 + \frac{\sigma_{13}I_p}{h\nu_p}n_1 + \frac{\sigma_{12}I_s}{h\nu_s}n_1 - \frac{\sigma_{21}I_s}{h\nu_s}n_2$$

$$\frac{\partial n1}{\partial t} = A_{21}n_2 - \frac{\sigma_{13}I_s}{h\nu_p}n_1 - \frac{\sigma_{12}I_s}{h\nu_s} - \frac{\sigma_{21}I_s}{h\nu_s},$$

where of course one of the equations is redundant, as the sum of the two level populations must stay unity in any case.

Various circumstances can lead to additional complications:

- It is common to insert additional terms, e.g. for energy transfer processes. There may be, e.g., terms proportional to n_2^2 for cooperative up conversion processes, where one ion in level 2 transfers energy to another ion in the same level, resulting in one more ion in the ground-state manifold and the other one in a higher level. Assuming that the higher levels quickly decay to level 2, the additions to the equation system are relatively simple.
- It is also possible to include energy transfers between different species of ions. A common case is that of erbium–ytterbium-doped fibers, where primarily ytterbium ions absorb pump radiation, and transfer energy to erbium ions. The model then includes level populations for all involved types of ions, and the corresponding coupling terms.
- Additional optical wavelengths can be involved e.g. if amplified spontaneous emission (ASE) occurs in a fiber amplifier. In numerical models, the ASE spectrum is divided into discrete wavelength slots, each one being associated with different values of the absorption and emission cross sections.
- Generally, the rate equations have to be solved for different positions within the gain medium, as the optical intensities depend on both the longitudinal and transverse coordinates.

Solving and Using the Rate Equations

For a given location in the gain medium and for given optical intensities, the temporal evolution of the population densities can be calculated by temporal integration of the rate equations. This can be done e.g. with the Runge–Kutta method. Note, however, that rate equations are often a so-called "stiff" system of differential equations, involving very different time constants. The fastest processes in the system (e.g. related to relatively fast non-radiative transitions) then force one to use a fairly small temporal step size, as the numerical solution would otherwise become unstable. At the same time, much slower processes often imply that the evolution must be calculated over a long time span, so that in effect many numerical steps are needed. For such reasons, it can be inefficient to calculate the steady-state population densities for given optical intensities simply by simulating the temporal evolution.

For simple cases, it is easy to calculate analytically the steady-state populations for given pump and signal intensities. For example, the simplified equation system above leads to the following result:

$$n_2 = \frac{\dfrac{\sigma_{13} I_p}{h\nu_p} + \dfrac{\sigma_{12} I_s}{h\nu_s}}{A_{21} + \dfrac{\sigma_{13} I_p}{h\nu_p} + \dfrac{(\sigma_{12} + \sigma_{21}) I_s}{h\nu_s}}$$

It is also possible to calculate the evolution of both the population densities and the optical powers, e.g., in order to model the laser dynamics of a Q-switched laser. The set of differential equations then includes the rate equations for the populations as well as dynamical equations for the optical powers.

In many situations, it is possible and convenient to derive equations for spatially averaged population densities. In the case of a simple level scheme, the population of the gain medium is then described with a single variable, which is directly related to the laser gain. This variable can then be used e.g. as a dynamic variable in the dynamic equations for population (or gain) and laser intensity.

Limitations of Rate Equation Modeling

Rate equations describe the statistical evolution of level populations, and are based on a number of assumptions. They do not describe coherent phenomena such as Rabi oscillations, as they average over many ions, which experience slightly different microscopic environments and different optical intensities. It is further assumed that all ions essentially function in the same way. The latter assumption can be violated e.g. if clustering occurs in the gain medium. In such a case, ions within clusters may exhibit, e.g., much stronger up conversion processes than other ions do, and should therefore be treated in the model as a separate species. Such extended models have been developed, but they are more complicated, and often involve a number of parameters which are hard to access experimentally.

Population Inversion

Population inversion is an important concept for lasers optics, as it is essential for light amplification by stimulated emission. The principle is general and can be applied for other types of lasers, amplifiers, and super luminescent light sources (e.g. semiconductor laser and SLD).

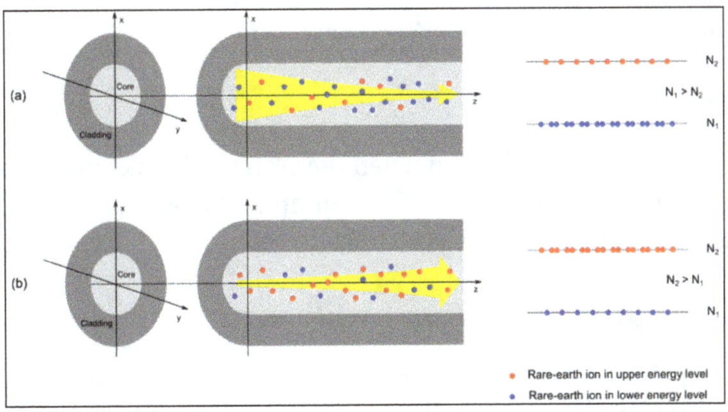

Schematic of population inversion and amplification of light in rare-earth-doped fiber. (a) Light absorption without population inversion, and (b) light amplification with population inversion.

Figure above schematically shows light absorption and amplification in a rare-earth-doped fiber. The fibers shown in Figure (a) and (b) have the same geometry, and are doped with the same amount of rare-earth elements; however the number of ions in the two energy states is different. In Figure (a), the number of ions in the lower energy state (N1) are smaller than that of the upper energy state (N2); i.e, N1>N2. On the other hand, in Figure (b), N2>N1. Roughly speaking, these two situations correspond to (a) weak pumping, and (b) strong pumping.

If the pumping is not enough and population inversion is not created, the absorption of light (by the ion in the lower energy level) surpasses the amplification and thus the light is attenuated. On the other hand, if population inversion is created, amplification by stimulated emission (by the ions in the upper energy state) surpasses absorption and thus the light is amplified.

Methods of Achieving Population Inversion

Under normal conditions, more electrons are in a lower energy state than in a higher energy state. Population inversion is a process of achieving more electrons in the higher energy state than the lower energy state.

In order to achieve population inversion, we need to supply energy to the laser medium. The process of supplying energy to the laser medium is called pumping. The source that supplies energy to the laser medium is called pump source. The type of pump source used is depends on the laser medium. Different pump sources are used for different laser mediums to achieve population inversion. Some of the most commonly used pump sources are as follows:

- Optical pumping,
- Electric discharge or excitation by electrons,
- Inelastic atom-atom collisions,
- Thermal pumping,
- Chemical reactions.

Population inversion is easily achieved when the system of molecules or atoms have the energy levels with favorable properties. For example, the upper energy level has a long lifetime and the lower energy level has a short lifetime.

Electric Discharge or Excitation by Electrons

Electric discharge refers to flow of electrons or electric current through a gas, liquid or solid. In this method of pumping, electric discharge acts as the pump source or energy source. A high voltage electric discharge (flow of electrons, electric charge, or electric current) is passed through the laser medium or gas. The intense electric field accelerates the electrons to high speeds and they collide with neutral atoms in the gas. As a result, the electrons in the lower energy state gains sufficient energy from external electrons and jumps into the higher energy state. This method of pumping is used in gas lasers such as argon lasers.

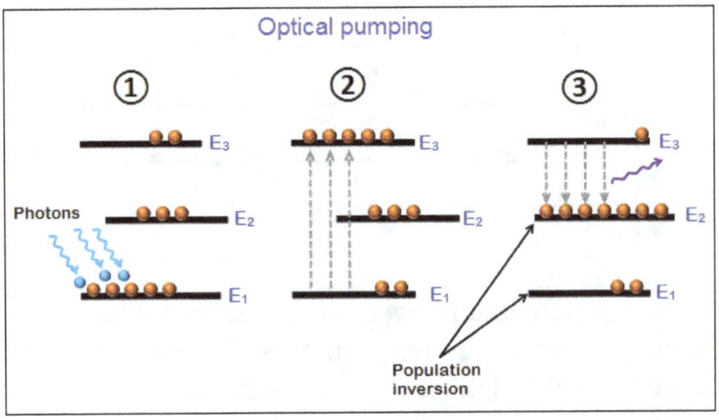

The process of achieving population inversion in the gas laser is almost similar to the solid laser. The only difference is the pump source used for supplying energy and the type of material or medium (solid or gas) used as a laser medium. In solid lasers, an external light source like xenon flash lamp is used as pump source whereas, in gas lasers, a high voltage electric discharge is used as a pump source.

Inelastic Atom-Atom Collisions

Like the electric discharge method, here also a high voltage electric discharge acts as a pump source. However, in this method, a combination of two types of gases, say X and Y are used. The excited state of gas X is represented as X^+ whereas gas Y is represented as Y^+. Both X and Y gases have the same excited states (X^+ and Y^+).

When high voltage electric discharge passes through a laser medium having two types of gases X and Y, the lower energy state electrons in gas X will move to the excited state X^+ similarly the lower energy state electrons in gas Y moves to the excited state Y^+.

Initially, during electric discharge, the lower energy state electrons in gas X or atom X gets excited to X^+ due to continuous collision with electrons. The excited state electrons in gas X^+ now collide with the lower energy state electrons in gas Y. As a result, the lower energy state electrons in gas Y gains sufficient energy and jump into the excited state Y^+. This method is used in the Helium–Neon (He-Ne) laser.

Thermal Pumping

Sometimes we can achieve population inversion by heating the laser medium. In thermal pumping, heat acts as the pump source or energy source. In this method, population inversion is achieved by supplying heat into the laser medium. When heat energy is supplied to the laser medium, the lower energy state electrons gains sufficient energy and jumps into the higher energy level.

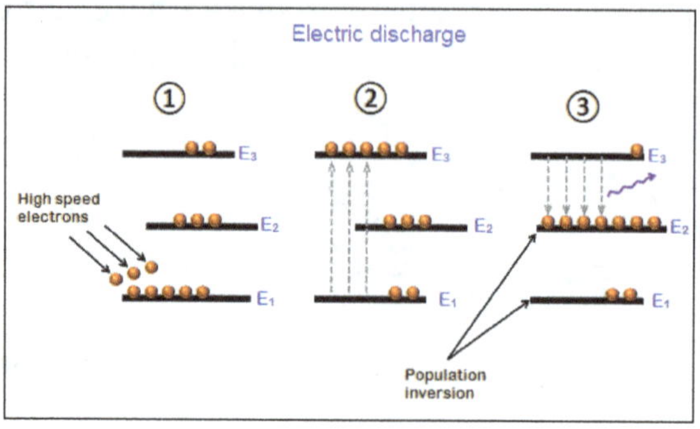

The process of achieving population inversion in thermal pumping is almost similar to the optical pumping or electric discharge method, except that in this method heat is used as pump source instead of light or electric discharge.

Chemical Reactions

If an atom or a molecule is produced through some chemical reaction and remains in an excited state at the time of production, then it can be used for pumping. The hydrogen fluoride molecule is produced in an excited state when hydrogen and fluorine gas chemically combine. The number of produced excited atoms or molecules is greater than the number of normal state atoms or molecules. Thus, population inversion is achieved. For example, $H_2 + F_2 \rightarrow 2HF$, in this chemical reaction, hydrogen (H_2) and fluorine (F_2) molecules are chemically combined to produce hydrogen fluoride molecule (2HF) in an excited state.

Optical Pumping

Optically pumping some medium essentially means to inject light in order to electronically excite the medium or some of its constituents into other (usually higher-lying) energy levels. In the context of lasers or laser amplifiers, the goal is to achieve a population inversion in the gain medium and thus to obtain optical amplification via stimulated emission for some range of optical frequencies. (The width of that range is called the gain bandwidth.) In other cases, such as in spectroscopic measurements, the goal can be to selectively populate a specific electronic level (e.g. some hyperfine sublevel), which does not necessarily have an energy well above the ground state.

Optical pumping processes can often be described with rate equation modeling. However, this disregards some aspects of the quantum nature of the atom–photon interaction. More comprehensive physical models exist which can also describe coherent phenomena such as Rabi oscillations.

Optically Pumped Lasers

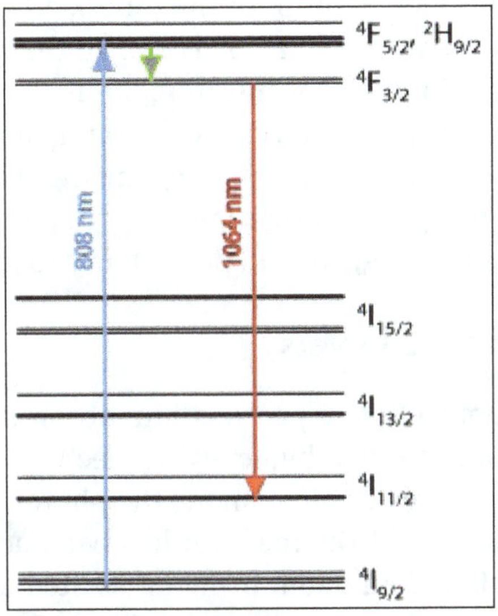

Neodymium ions (e.g. in a YAG laser) can be optically pumped to some high-lying energy levels, from where they quickly decay (green arrow) into the upper laser level, from where stimulated emission at the laser wavelength of 1064 nm (for example) can occur.

Various types of lasers can be optically pumped:

- The most common optically pumped lasers are doped-insulator solid-state lasers. As the host medium (a laser crystal, glass or piece of ceramic) is electrically insulating, optical pumping is the only way to supply the laser-active ions (e.g. rare earth ions) with energy.
- Semiconductor lasers are in most cases electrically pumped, but optical pumping can have certain advantages. In the case of vertical external cavity surface-emitting lasers (VECSELs) it allows for a laser output with very high brightness (high power combined with high beam quality). In other cases, optical pumping is sometimes used for testing purposes, e.g. when new semiconductor compositions are investigated. Optically pumped semiconductor gain media generally have a simpler structure than their electrically pumped cousins.
- In rare cases, gas lasers (particularly alkali vapor lasers) are optically pumped.

Common types of optical pump sources are:

- Discharge lamps (→ lamp-pumped lasers).
- Laser diodes (→ diode-pumped lasers).
- Other types of lasers or laser sources.

Examples of the latter case are titanium–sapphire lasers pumped with frequency-doubled solid-state lasers, and dye lasers pumped with gas lasers. A requirement for achieving high power conversion efficiency in an optically pumped laser is that the pump light is efficiently absorbed in the gain medium. This can be the case if the gain medium is sufficiently long, has a high doping concentration, and a sufficiently wide optical frequency range for the pump light. In some cases, the pump absorption efficiency can be increased by arranging for multiple passes of the pump light through the gain medium. This technique is often used e.g. in thin-disk lasers. A frequently used alternative to optical pumping is electrical pumping, applied particularly to laser diodes and gas lasers.

Using Multiple Electronic Levels

In the simplest case, the optically pumped medium absorbs light, and each absorbed photon excites one atom or ion into a higher-lying electronic level. The energy difference of the involved electronic levels must match the photon energy of the pump light. Once some degree of excitation of the medium has been achieved, the same kind of light can also cause stimulated emission, bringing excited atoms or ions back to the lower level. For that reason, a medium with a simple two-level scheme cannot reach a population inversion.

Optical pumping in lasers involves at least three different energy levels. In the simplest case, laser-active ions are pumped from their ground state to a higher-lying level, from where they undergo a quick radiative or non-radiative decay into an excited level with somewhat lower energy. As pump light cannot cause stimulated emission from this level down to the ground state (due to the too high photon energy), a population inversion can be achieved for a sufficiently high pump intensity. By introducing a fourth energy level, one can obtain a four-level laser system, where the laser transition involves two intermediate levels. If the lower level population is kept small due to a quick decay to the ground state, population inversion can be achieved even when only a tiny fraction of the ions is excited. The article on four-level and three-level gain media explains the details.

In-Band Pumping

Solid state laser gain media often exhibit slightly non-degenerate Stark level manifolds. Due to the energy variations within each manifold, optically pumped laser operation is possible even with only two involved manifolds: ions are pumped from the lower manifold (usually the ground-state manifold) to some higher manifold, and the laser transition directly leads from there back to the lower manifold, with no intermediate manifolds. This pump scheme, called in-band pumping, can be used with various laser-active ions:

- Ytterbium-doped gain media exhibit only two Stark level manifolds, so there is no other option than in-band pumping.
- Erbium-doped gain media can be in-band-pumped, using the lowest two levels ($^4I_{15/2}$ and $^4I_{13/2}$). This means that e.g. an erbium-doped fiber amplifier is pumped

around 1450 nm while amplifying around 1550 nm. Similarly, there are Er:YAG lasers which are pumped with an erbium-doped fiber laser or amplifier, and emitting at 1645 nm.

- The same scheme can also be applied e.g. to thulium-doped media.

The term in-band pumping is also often used in cases where the laser-active ions are pumped from the ground-state manifold (not being the lower laser level) directly into the upper laser level. For example, this happens when a neodymium-doped laser crystal is pumped around 0.88 μm.

In-band pumping often leads to a small quantum defect, but also often to significant effects of reabsorption from the lower laser level (→ quasi-three level behavior). Also, the achievable degree of excitation (and thus the laser gain) can be limited by stimulated emission caused by the pump light. A somewhat shorter pump wavelength may mitigate the latter problem, but this can decrease the pump absorption efficiency.

Side Pumping Versus end Pumping

In an optically pumped laser, pump light can be injected into the gain medium from different directions. Side pumping means that the light is injected in directions which are roughly perpendicular to that of the laser beam. In the case of end pumping, the pump light is approximately collinear with the laser beam. The pump geometry has implications for the required beam quality, but also the achieved power conversion efficiency, the laser gain, and the laser beam quality.

Requirements on the Pump Light

Pump light for optical pumping has to fulfill a number of requirements:

- The optical spectrum of the pump light must be suitable. Ideally, all the photons should have a suitable energy for the wanted electronic transitions. However, certain laser-active ions (e.g. neodymium ions) can also be pumped with fairly broadband light e.g. for flash lamps or arc lamps, albeit with a strongly reduced power conversion efficiency.

- The pump intensity must be sufficiently high. Lasers are often pumped with intensities of the order of the saturation intensity of the laser transition, but four-level lasers can also be operated with lower pump intensities.

- Depending on the geometry, there can be more or less stringent requirements on the pump beam quality. This applies mostly to end-pumped lasers.

- In some cases, the polarization state of the pump light is also important. Some non-isotropic gain media, such as Nd:YVO_4, exhibit very different levels of absorption for different polarization directions. In spectroscopy, circularly polarized light is sometimes required for populating certain hyperfine levels.

- The intensity noise of the pump source should not be too large, because at least its low-frequency components can be transferred to the laser output.

Optical Pumping in Fundamental Physics

Techniques of optical pumping are also used outside the area of laser physics, often in fundamental research. For example, various methods of laser cooling essentially depend on optical pumping of atoms or ions, for example suspended in an optical trap. In many cases, certain selection rules play a vital role, which reflect principles like the conservation of angular momentum: the absorption of photons of circularly polarized light must accordingly change the angular momentum of the atoms, i.e., it must lead to final states having correspondingly modified angular momentum. Therefore, it is possible, for example, to selectively pump certain states within a manifold of states having essentially the same excitation energy. In many cases, such processes involve multiple steps of optical pumping and spontaneous emission until the final target states are largely populated.

Interestingly, optical pumping can reduce the entropy of atoms or ions; such processes are associated with an increase of entropy of the light field, such that the total entropy is never decreased and in fact often increased. Optical pumping can also be utilized for isotope separation, because different isotopes of a substance exhibit slightly different transition energies.

Laser in Two Level Systems

Let's first consider a laser medium whose atoms have only two energy states: a ground state and one excited state. In such an idealized atom the only possible transitions are excitation from the ground state to the excited state, and de-excitation from the excited state back into ground state. Could such an atom be used to make a laser?

There are several important conditions that our laser must satisfy. First of all, the light that it produces must be coherent. That is to say, it must emit photons that are in-phase with one another. Secondly, it should emit monochromatic light, i.e. photons of the same frequency (or wavelength).Thirdly, it would be desirable if our laser's output were collimated, producing a sharply defined "pencil-like" beam of light (this is not crucial, but clearly a desirable condition). Lastly, it would also be desirable for our laser to be efficient, i.e. the higher the ratio of output energy - to - input energy, the better.

Let us begin by examining the requirements for our first condition for lasing, coherence. This condition is satisfied only when the lasing transition occurs through stimulated emission. As we have already seen, stimulated emission produces identical photons that are of equal energy and phase and travel in the same direction. But for stimulated emission to take place a "passer-by" photon whose energy is just equal to the de-excitation energy must approach the excited atom before it de-excites via spontaneous emission. Typically, a photon emitted by the spontaneous emission serves as the seed to trigger a collection of stimulated emissions. Still, if the lifetime of the excited state is too short, then there will not be enough excited atoms around to undergo stimulated emission. So, the first criteria that we need to satisfy is that the upper lasing state must have a relatively long lifetime, otherwise known as a meta-stable state, with typical lifetimes in the milliseconds range. In addition to the requirement of a long lifetime, we need to ensure that the likelihood of absorption of the "passer-by" photons is minimized. This likelihood is directly related to the ratio of the atoms in their ground state versus those in the excited state. The smaller this ratio, the more likely that the "passer-by" photon will cause a stimulated emission rather than get absorbed. So, to satisfy this requirement, we need to produce a population inversion: create more atoms in the excited state than those in the ground state.

Achieving population inversion in a two-level atom is not very practical. Such a task would require a very strong pumping transition that would send any decaying atom back into its excited state. This would be similar to reversing the flow of water in a water fall. It can be done, but is very energy costly and inefficient. In a sense, the pumping transition would have to work against the lasing transition.

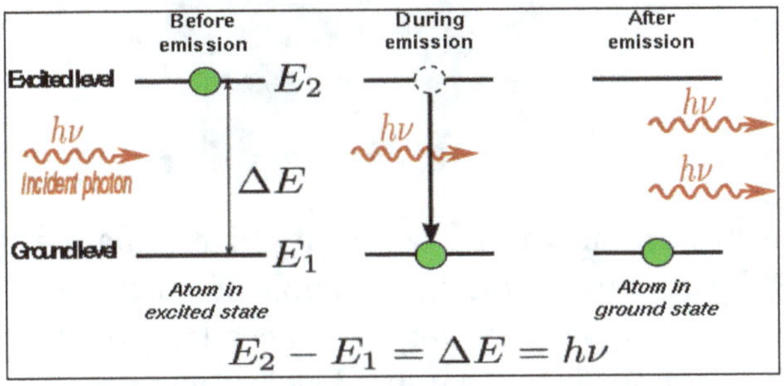

Stimulated emission in a two-level transition.

It is clear, from the above diagram that in the two-level atom the pump is, in a way, the laser itself! Such a two-level laser would work only in jolts. That is to say, once the population inversion is achieved the laser would lase. But immediately it would end up with more atoms in the lower level. Such two-level lasers involve a more complicated process.

Laser in Three Level Systems

For optical frequencies, population inversion cannot be achieved in a two level system. In 1956 Bloembergen proposed a mechanism in which atoms are pumped into an excited state u by an external source of energy (such as by an electric pulse or by optical illumination).

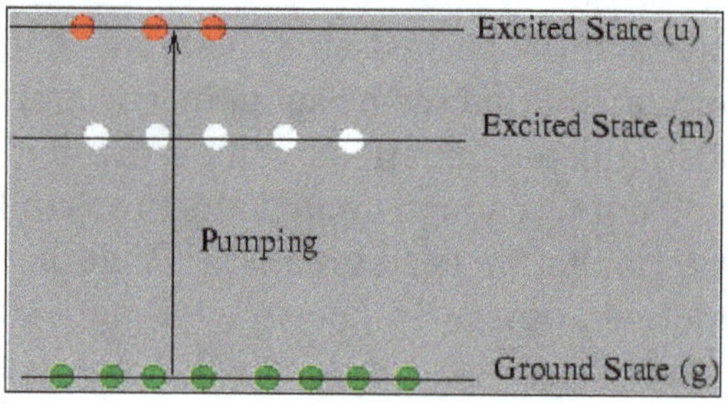

The system, in addition to the state u, has an excited state m which is a metastable state, i.e. a state in which the atom has a long life time. Atoms from the upper level u decays spontaneously to this metastable state m. Life time in the level is such that the rate of spontaneous decay from level to the ground level g is slower than the rate at which atoms decay from u to m. This results in a population inversion between the metastable level and the ground state.

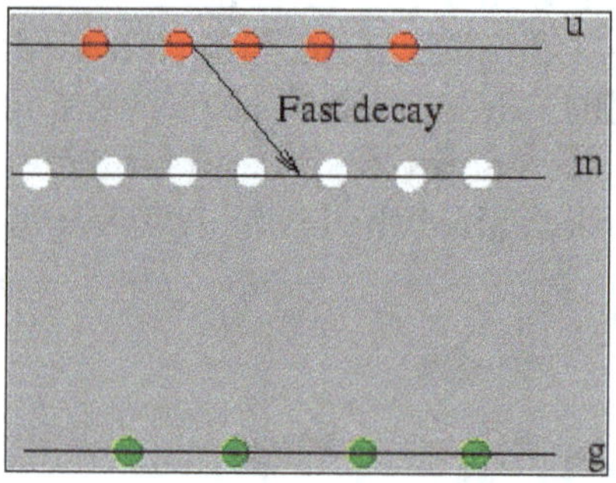

The emitted photons are confined to a laser cavity to stimulate further emission from the excited atoms. Ruby laser works on the principle of a three level system. The pumping power required for such a system is very high because more than half of the ground state atoms have to be pumped into the upper level to achieve population inversion.

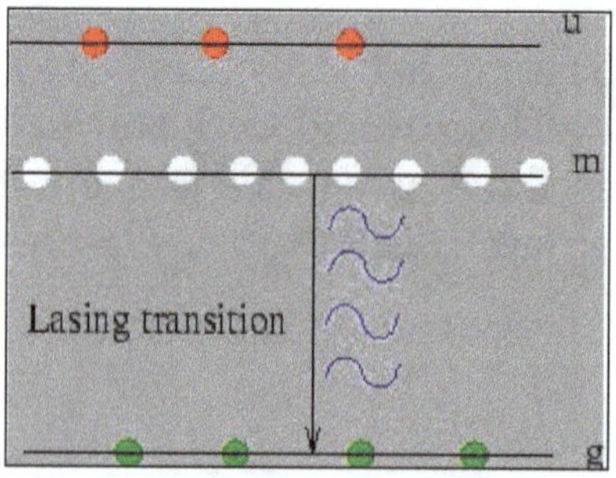

A few additional points to be noted about the three level system are:

- As $N_m \simeq N_g = N/2$ and $N_m - N_g \ll N$, once the population inversion is achieved, the power required to maintain it is small.

- $u \rightarrow m$ transition is generally radiation less, the energy being given away to the lattice.
- As more than half of the atoms are to be raised to the level m, the probability of spontaneous emission is also much higher.
- Laser transition occurs from the metastable state to the lowest possible states which are well separated. This leads to low efficiency.

Rate Equations for a Three Level System

In a three level system, the laser transition is from the metastable state m to the ground state g. In the following analysis, we will ignore the effects of degeneracy. As in the case of two level system, we denote by W_P the transition rate induced by pumping from the ground state (g) to the top level m or vice versa. W_P is clearly proportional to the pumping rate. As the metastable state is long lived, we assume that $T_{u,m}$ is much smaller than $T_{u,g}$ either or $T_{m,g}$, where $T_{i,j}$ denotes the lifetime of transition $i \rightarrow j$. As a result, the population of the upper level is nearly zero, and correspondingly:

$$N = N_{u+} N_m + N_g \simeq N_m + N_g$$

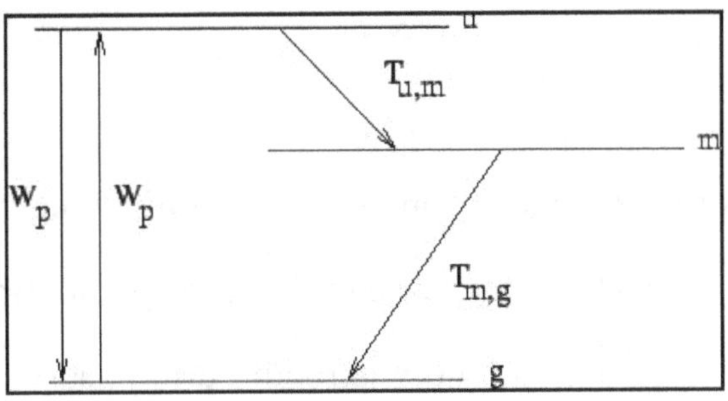

The rate equation for the level u may be written as:

$$\frac{dN_u}{dt} = W_p\left(N_g - N_u\right) - \frac{N_u}{T_{u,m}}$$

The equation for the metastable level m is,

$$\frac{dN_u}{dt} = \frac{N_u}{T_{u,m}} - \frac{N_m}{T_{m,g}}$$

The rate equation for the ground state is obtained from the above two equations by number conservation:

$$\frac{dN_g}{dt} = \frac{d}{dt}(N_u + N_m) = -W_p\left(N_g - N_u\right) + \frac{N_m}{T_{m,g}}$$

In the steady state, each of the time derivatives is zero, which gives:

$$N_u = \frac{W_p T_{m,g}}{W_p T_{u,m} + 1} N_g$$

$$N_m = \frac{W_p T_{m,g}}{W_p T_{u,m} + 1} N_g$$

The equation for population inversion becomes:

$$\frac{\Delta N}{N} = \frac{N_m - N_u}{N} = \frac{W_p T_{m,g}(1-\beta) - 1}{W_p T_{m,g}(1+2\beta) + 1}$$

Where,

$$\beta = \frac{N_u}{N_m} = \frac{T_{u,m}}{T_{m,g}}$$

For population inversion to occur, the numerator of $\frac{\Delta N}{N} = \frac{N_m - N_u}{N} = \frac{W_p T_{m,g}(1-\beta) - 1}{W_p T_{m,g}(1+2\beta) + 1}$ must be positive,

$W_p T_{m,g}(1-\beta) > 1$ i.e.,

$$W_p T_{m,g} > \frac{1}{1-\beta}$$

From $\beta = \frac{N_u}{N_m} = \frac{T_{u,m}}{T_{m,g}}$, as the lifetime for $m \rightarrow g$ transition is longer than that for $u \rightarrow m$ transition. Thus $W_p T_{m,g} > 1$ is a sufficient condition for population inversion. A minimum pumping power determined by (C) is required to ensure population inversion. For $\beta = 0$,

$$\frac{\Delta N}{N} = \frac{W_p T_{m,g} - 1}{W_p T_{m,g} + 1}$$

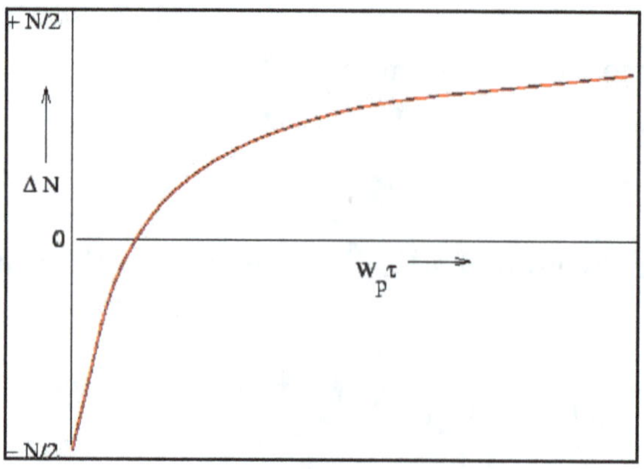

As N_g is ground level, its value is initially very large and at least $N/2$ atoms must be transferred to the upper level achieve population inversion. Hence a three level system is not very energy efficient.

Laser in Four Level Systems

One of the most popular and low cost lasers is helium-neon laser, which works on the basis of a four level system with two levels intermediate between the ground state g and the pumping level p. The ground state atoms are electrically pumped to a short lived state p. Atoms from this state undergo fast decay to a metastable state m. Between m and g, yet another short lived excited state i exists. A population inversion takes place between this intermediate state i and the metastable state m, between which the lasing transition occurs.

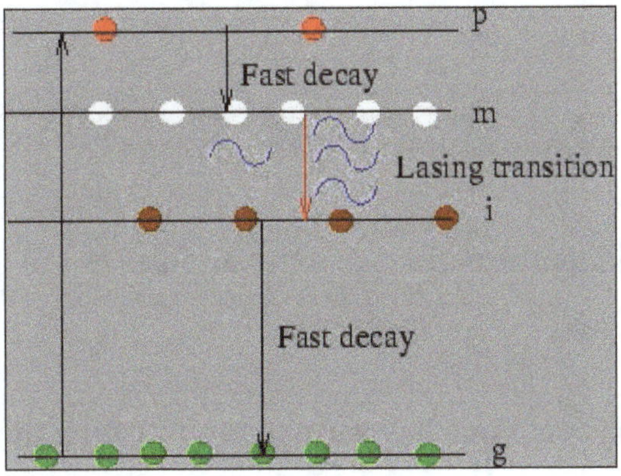

Rate Equations for a Four Level System

We denote the ground state as level 1 and the pumping level as level 4. Lasing transition occurs between levels 2 and 3 (2→3). Let $W_p = W_{1\to4} = W_{4\to1}$ be pumping induced transition rate between the ground level and the uppermost level. The lifetime of the lasing transition $T_{3,2}$ is much greater than all other lifetimes in the problem.

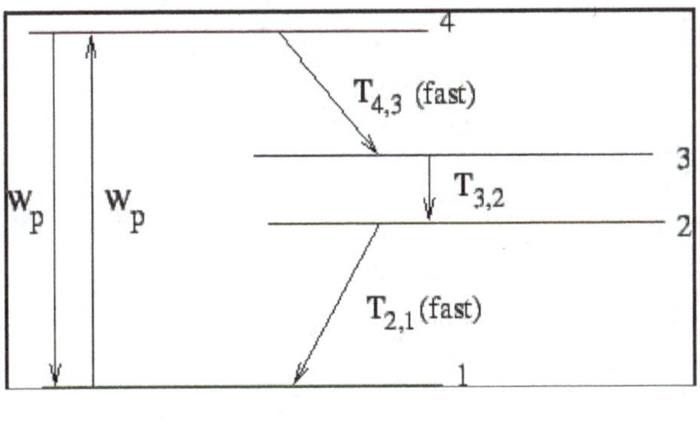

Let T be the lifetime of the level 4 for decay into any of the lower states. As the decay constants are additive, we have:

$$\frac{1}{T_4}=\frac{1}{T_{4,1}}+\frac{1}{T_{4,2}}+\frac{1}{T_{4,3}}$$

A similar convention will be used to denote the lifetime of a state against decay to any of the lower levels. Thus,

$$\frac{1}{T_3}=\frac{1}{T_{3,2}}+\frac{1}{T_{3,1}}$$

The rate equations may now be written as follows:

$$\frac{dN_4}{dt}=W_p\left(N_1-N_4\right)-\frac{N_4}{T_4}$$

$$\frac{dN_3}{dt}=\frac{N_4}{T_{4,3}}-\frac{N_3}{T_3}$$

$$\frac{dN_2}{dt}=\frac{N_4}{T_{4,2}}-\frac{N_3}{T_{3,2}}-\frac{N_2}{T_{2,1}}$$

These are three independent equations in four variables as the total number:

$$N=N_{1+}N_2+N_3+N_4$$

We will now solve these equations in steady state by putting the time derivatives to be zero. From equations $\frac{dN_4}{dt}=W_p\left(N_1-N_4\right)-\frac{N_4}{T_4}$, we get,

$$N_4=\frac{W_P T_4}{1+W_P T_4}N_1$$

From equation $\frac{dN_3}{dt}=\frac{N_4}{T_{4,3}}-\frac{N_3}{T_3}$, we get,

$$N_3=\frac{T_3}{T_{4,3}}N_4$$

Equations $\frac{dN_2}{dt}=\frac{N_4}{T_{4,2}}-\frac{N_3}{T_{3,2}}-\frac{N_2}{T_{2,1}}$ gives,

$$N_2=\frac{T_{2,1}}{T_{4,2}}N_4+\frac{T_{2,1}}{T_{3,2}}N_3$$

$$=\left(\frac{T_{2,1}}{T_{4,2}}\frac{T_{4,3}}{T_3}+\frac{T_{2,1}}{T_{3,2}}\right)N_3\equiv\beta N_3$$

Where,

$$\left(\frac{T_{2,1}}{T_{4,2}}\frac{T_{4,3}}{T_3}+\frac{T_{2,1}}{T_{3,2}}\right)$$

If $\beta<1$, there will be more population in the upper laser level than in the lower one, leading to a population inversion. Using the above, and after straight forward simplification, we get:

$$N=\frac{1+2W_pT_4+(1+\beta)W_p\frac{T_4T_3}{T_4T_3}}{1+W_pT_4}N_1$$

And the population difference $N_3-N_2=\Delta N$ given by,

$$\Delta N=(1-\beta)\frac{T_3}{T_{4,3}}\frac{W_pT_4}{1+W_pT_4}N_1$$

Combining the above two equation:

$$\frac{\Delta N}{N}=\frac{(1-\beta)W_PT_4}{1+\left[(1+\beta)+2\frac{T_{4,3}}{T_3}\right]W_P\frac{T_4T_3}{T_{4,3}}}$$

Let us define Quantum Efficiency η as the fraction of atoms excited from the ground state $\frac{dN_4}{dt}=W_p(N_1-N_4)-\frac{N_4}{T_4}$ which ultimately results in stimulated emission. is thus a product of the fraction which arrives at the upper laser level $\frac{dN_2}{dt}=\frac{N_4}{T_{4,2}}-\frac{N_3}{T_{3,2}}-\frac{N_2}{T_{2,1}}$ and the fraction of atoms in the upper laser level which make radiative transition to the lower laser level $\frac{dN_3}{dt}=\frac{N_4}{T_{4,3}}-\frac{N_3}{T_3}$.

Fraction of atoms in level 4 which arrive at level 3 is given by the ratio $T_4/T_{4,3}$ and the fraction of atoms in level 3 which radiatively make a transition to level 2 is $T_3/T_{3,2}^{rad}$. We have used a superscript rad to indicate that only the radiative component of the transition from 3 to 2 is considered. Thus

$$\eta=\frac{T_4}{T_{4,3}}\frac{T_3}{T_{3,2}^{rad}}$$

Substituting this in the previous equation, we get, for $\Delta N/N$

$$\frac{\Delta N}{N}=\frac{(1-\beta)W_p\eta T_{3,2}^{rad}}{1+\left[(1+\beta)+2\frac{T_{4,3}}{T_3}\right]W_p\eta T_{3,2}^{rad}}$$

To ensure that most of the atoms excited by pumping participate in laser transition, the life time in the level 3 must be the longest. Using $T_{4,3} \ll T_3$, we may ignore this term in the denominator of the above. Further, $T_{3,2} \gg T_{2,1}$. Using these, one can see that is small and approaches zero.

Thus ΔN/N remains positive and population inversion occurs even for very small pumping power. For $\beta \to 0$, the expression for ΔN/N is given by,

$$\frac{\Delta N}{N} = \frac{W_p \eta T_{3,2}^{rad}}{1 + W_p \eta T_{3,2}^{rad}}$$

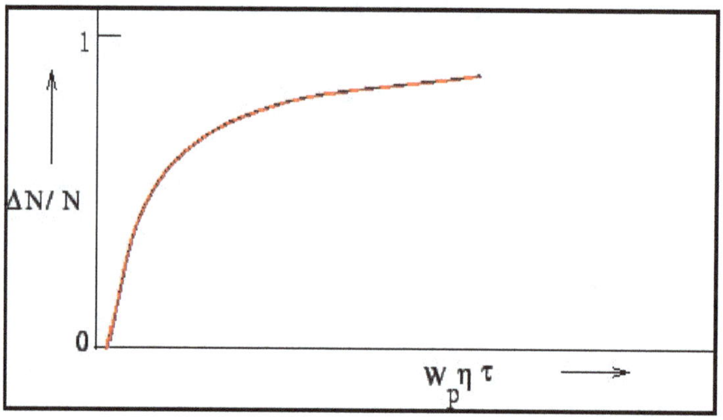

References

- Plancks-quantum-theory: lumenlearning.com, Retrieved 16, January 2020
- Einstein-coefficient-relations: winnerscience.com, Retrieved 30, May 2020
- Rate-equation-modeling: rp-photonics.com, Retrieved 05, February 2020
- Population-inversion: fiberlabs.com, Retrieved 10, April 2020
- Optical-pumping: rp-photonics.com, Retrieved 17, August 2020

3

Laser Machining

Machining involves cutting a material into a desired final shape and size. Laser machining is a thermal machining process wherein a laser beam is used for the production of heat which removes material from a workpiece. The topics elaborated in this chapter will help in gaining a better perspective about laser machining as well as laser beam welding and laser shock peening.

LM is a method of cutting metal or refractory materials by melting and vaporizing the material with an intense beam of light from a laser. Drilling by laser, although costly in energy since material must be melted and vaporized to be removed, is used to cut small holes (0.005 to 0.05 inch [0.13 to 1.3 millimetres]) in materials that are too difficult to machine by traditional methods. A common application is the laser drilling of diamonds to be used as dies for drawing wire. Lasers also are used to drill and cut ceramics and substrates for integrated circuits; the aircraft industry uses CNC-controlled lasers to cut profiles and to drill holes in engine parts.

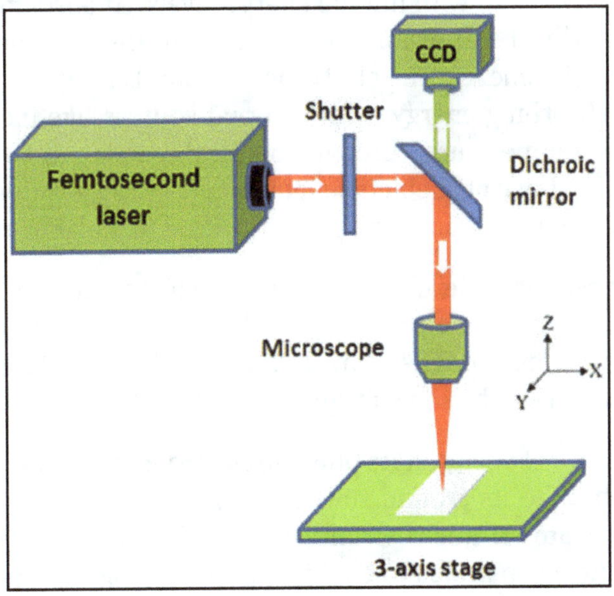

Laser Beam Machining

Laser Beam Machining or more broadly laser material processing deals with machining and material processing like heat treatment, alloying, cladding, sheet metal bending etc. Such processing is carried out utilizing the energy of coherent photons or laser beam, which is mostly converted into thermal energy upon interaction with most of the materials. Nowadays, laser is also finding application in regenerative machining or rapid prototyping as in processes like stereo-lithography, selective laser sintering etc.

Laser beam can very easily be focused using optical lenses as their wavelength ranges from half micron to around 70 microns. Focussed laser beam as indicated earlier can have power density in excess of 1 MW/mm^2. As laser interacts with the material, the energy of the photon is absorbed by the work material leading to rapid substantial rise in local temperature. This in turn results in melting and vaporisation of the work material and finally material removal.

Laser Beam Machining – The Lasing Process

Lasing process describes the basic operation of laser, i.e. generation of coherent (both temporal and spatial) beam of light by "light amplification" using "stimulated emission".

In the model of atom, negatively charged electrons rotate around the positively charged nucleus in some specified orbital paths. The geometry and radii of such orbital paths depend on a variety of parameters like number of electrons, presence of neighbouring atoms and their electron structure, presence of electromagnetic field etc. Each of the orbital electrons is associated with unique energy levels. At absolute zero temperature an atom is considered to be at ground level, when all the electrons occupy their respective lowest potential energy. The electrons at ground state can be excited to higher state of energy by absorbing energy form external sources like increase in electronic vibration at elevated temperature, through chemical reaction as well as via absorbing energy of the photon. Above fig. depicts schematically the absorption of a photon by an electron. The electron moves from a lower energy level to a higher energy level.

On reaching the higher energy level, the electron reaches an unstable energy band. And it comes back to its ground state within a very small time by releasing a photon. This is called spontaneous emission. The same is shown in figures below the spontaneously emitted photon would have the same frequency as that of the "exciting" photon.

Sometimes such change of energy state puts the electrons in a meta-stable energy band. Instead of coming back to its ground state immediately (within tens of ns) it stays at the elevated energy state for micro to milliseconds. In a material, if more number of electrons can be somehow pumped to the higher meta-stable energy state as compared to number of atoms at ground state, then it is called "population inversion". Such electrons, at higher energy meta-stable state, can return to the ground state in the form of an avalanche provided stimulated by a photon of suitable frequency or energy. This is called stimulated emission. Fig below shows one such higher state electron in meta-stable orbit. If it is stimulated by a photon of suitable energy then the electron will come down to the lower energy state and in turn one original photon, another emitted photon by stimulation having some temporal and spatial phase would be available. In this way coherent laser beam can be produced.

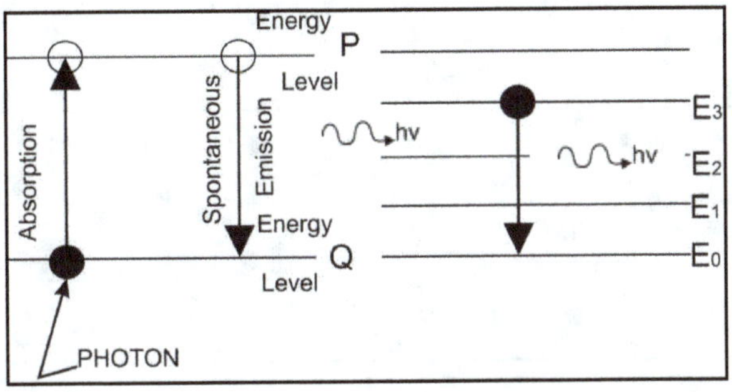

Fig: Energy bands in materials.

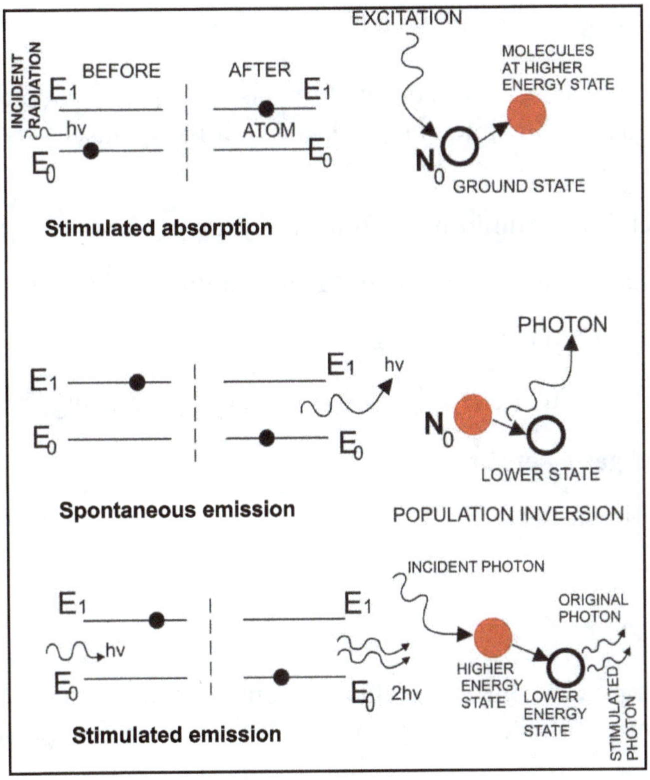

Fig: Spontaneous and stimulated emissions.

Figure below shows working of a laser. There is a gas in a cylindrical glass vessel. This gas is called the lasing medium. One end of the glass is blocked with a 100% reflective mirror and the other end is having a partially reflective mirror. Population inversion can be carried out by exciting the gas atoms or molecules by pumping it with flash lamps. Then stimulated emission would initiate lasing action. Stimulated emission of photons could be in all directions. Most of the stimulated photons, not along the longitudinal direction would be lost and generate waste heat. The photons in the longitudinal direction would form coherent, highly directional, intense laser beam.

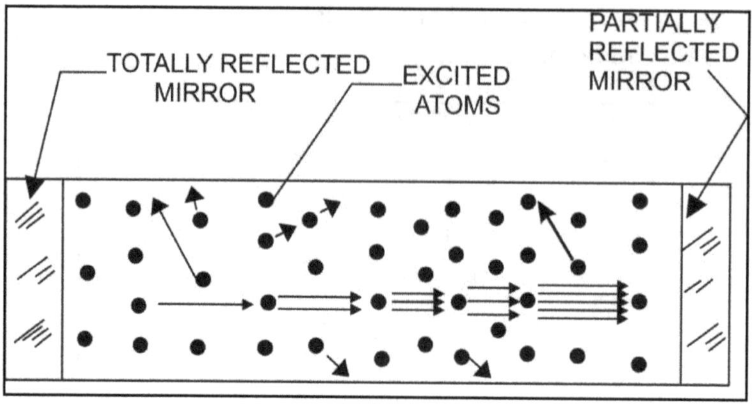

Fig: Lasing action.

Lasing Medium

Many materials can be used as the heart of the laser. Depending on the lasing medium lasers are classified as solid state and gas laser. Solid-state lasers are commonly of the following type:

- Ruby which is a chromium– alumina alloy having a wavelength of 0.7 μm.
- Nd- glass lasers having a wavelength of 1.64 μm.
- Nd- YAG laser having a wavelength of 1.06 μm.

These solid-state lasers are generally used in material processing.

The generally used gas lasers are:

- Helium – Neon;
- Argon;
- CO_2 etc.

Lasers can be operated in continuous mode or pulsed mode. Typically CO2 gas laser is operated in continuous mode and Nd – YAG laser is operated in pulsed mode.

Laser Construction

Figure below shows a typical Nd-YAG laser. Nd-YAG laser is pumped using flash tube. Flash tubes can be helical, as shown in Figure below, or they can be flat. Typically the lasing material is at the focal plane of the flash tube. Though helical flash tubes provide better pumping, they are difficult to maintain.

Figure above shows the electrical circuit for operation of a solid-state laser. The flash tube is operated in pulsed mode by charging and discharging of the capacitor. Thus the pulse on time is decided by the resistance on the flash tube side and pulse off time is decided by the charging resistance. There is also a high voltage switching supply for initiation of pulses.

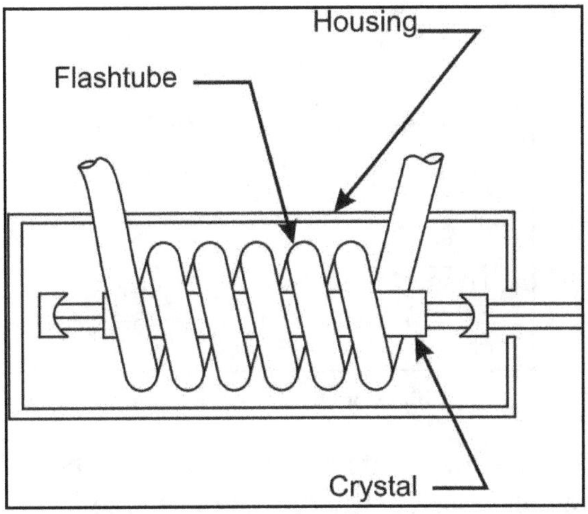

Fig: Solid-state laser with its optical pumping unit.

Figure below shows a CO_2 laser. Gas lasers can be axial flow, as shown in Figure below, transverse flow and folded axial flow as shown in figure below, The power of a CO_2 laser is typically around 100 Watt per metre of tube length. Thus to make a high power laser, a rather long tube is required which is quite inconvenient. For optimal use of floor space, high-powered CO_2 lasers are made of folded design.

In a CO_2 laser, a mixture of CO_2, N2 and He continuously circulate through the gas tube. Such continuous recirculation of gas is done to minimize.

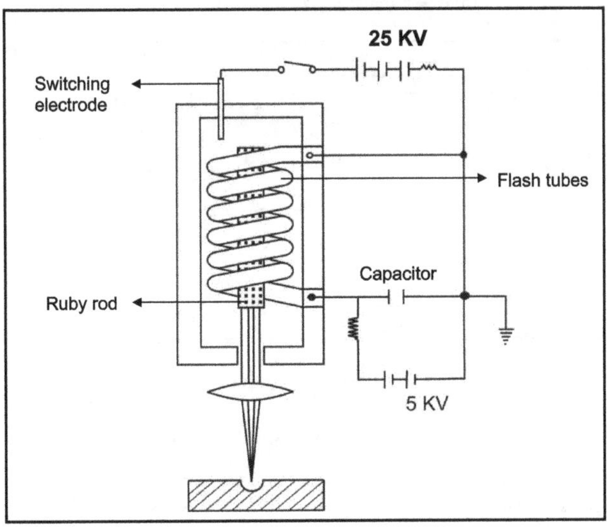

Fig: Working of a solid-state laser.

Consumption of gases. CO_2 acts as the main lasing medium whereas nitrogen helps in sustaining the gas plasma. Helium on the other hand helps in cooling the gases.

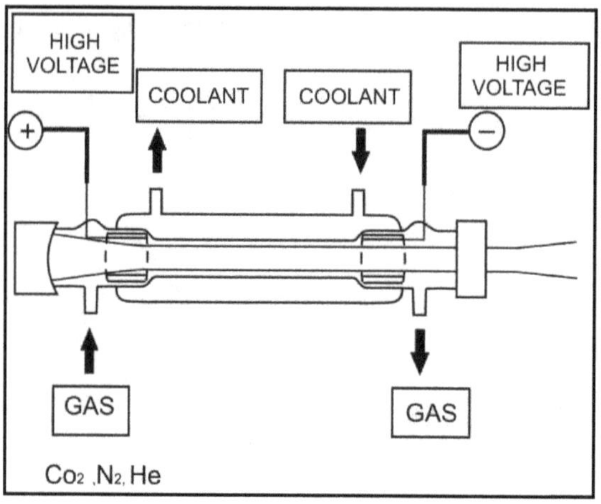

Fig: Construction of a CO_2 laser.

As shown in Figure above high voltage is applied at the two ends leading to discharge and formation of gas plasma. Energy of this discharge leads to population inversion and lasing action. At the two ends of the laser we have one 100% reflector and one partial reflector. The 100% reflector redirects the photons inside the gas tube and partial reflector allows a part of the laser beam to be issued so that the same can be used for material processing. Typically the laser tube is cooled externally as well.

As had been indicated earlier CO_2 lasers are folded to achieve high power. Figure below shows a similar folded axial flow laser. In folded laser there would be a few 100% reflective turning mirrors for manoeuvring the laser beam from gas supply as well as high voltage supply as shown in Figure below.

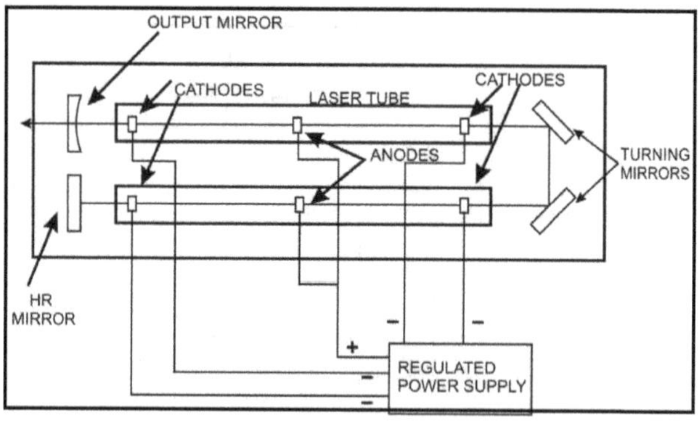

Fig: Construction of folded gas laser.

Table below shows the capability and characteristics of common lasers.

Application	Type of laser
Large holes upto 1.5 mm dia. Large holes (trepanned) Small holes > 0.25 mm dia. Drilling (punching or percussion)	Ruby, Nd-glass, Nd-YAG Nd-YAG, CO_2 Ruby, Nd-glass, Nd-YAG Nd-YAG, Ruby

Thick cutting Thin slitting of metals Thin slitting of plastics	CO_2 with gas assist Nd-YAG CO_2
Plastics Metals Organics, Non-metal Ceramics	CO_2 Nd-YAG, ruby, Nd-glass Pulsed CO_2 Pulsed CO_2, Nd-YAG

Table below process characteristics of different lasers:

Lasing materials	**Ruby**	**Nd-YAG**	**Nd-glass**	**CO_2**
Type	Solid state	Solid state	Solid state	Gas
Composition	0.03 – 0.7% Nd in $Al3O_2$	1% Nd doped Yttrium – Aluminium- Garnet	2-6% Nd in glass	CO_2+He+N_2 (3:8:4)
Wavelength (radiation)	0.69 μm	1.064 μm	1.064 μm	10.6 μm
Efficiency	1% max.	2%	2%	10-15%
Beam mode	Pulsed or CW	Pulsed or CW	Pulsed	Pulsed or CW
Spot size	0.015 mm	0.015 mm	0.025 mm	0.075 mm
Pulse repetition rate (normal operation).	1-10 pps	1-300 pps or CW	1-3 pps	CW
Beam output	10-100 W	10-1000 W	10 – 100 W	0.1 – 10 kW
Peak power	200 kW	400 kW	200 kW	100 kW

Application

Laser can be used in wide range of manufacturing applications:

- Material removal – drilling, cutting and tre-panning
- Welding;
- Cladding;
- Alloying.

Drilling micro-sized holes using laser in difficult – to – machine materials is the most dominant application in industry. In laser drilling the laser beam is focused over the desired spot size. For thin sheets pulse laser can be used. For thicker ones continuous laser may be used.

Advantages

- In laser machining there is no physical tool. Thus no machining force or wear of the tool takes place;
- Large aspect ratio in laser drilling can be achieved along with acceptable accuracy or dimension, form or location;

- Micro-holes can be drilled in difficult – to – machine materials;
- Though laser processing is a thermal processing but heat affected zone specially in pulse laser processing is not very significant due to shorter pulse duration.

Limitations

- High initial capital cost;
- High maintenance cost;
- Not very efficient process;
- Presence of Heat Affected Zone – specially in gas assist CO_2 laser cutting;
- Thermal process – not suitable for heat sensitive materials like aluminium glass fibre laminate.

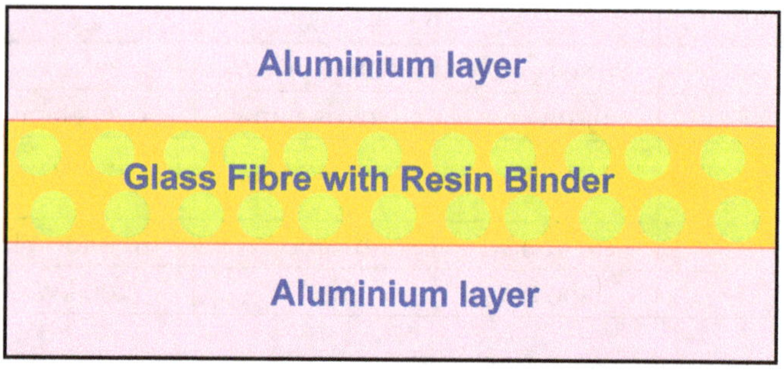

Fig: Aluminium Glass Fibre Laminate – heat sensitive glass fibre layer due to presence of resin as binder.

Laser Beam Welding

Laser beam welding is a fusion welding process in which two metal pieces are joined together by the use of laser. The laser beams are focused to the cavity between the two metal pieces to be joined. The laser beams have enough energy and when it strikes the metal pieces produces heat that melts the material from the two metal pieces and fills the cavity. After cooling a strong weld is formed between the two pieces.

Working Principle

It works on the principle that when electrons of an atom gets excited by absorbing some energy. And then after some time when it returns back to its ground state, it emits a photon of light. The concentration of this emitted photon increased by stimulated emission of radiation and we get a high energy concentrated laser beam.

Light amplification by stimulated emission of radiation is called laser.

Main Parts

The main parts or equipment of laser beam welding are:

1. Laser machine: It is a machine that is used to produce laser for welding. The main components of laser machine are shown below.
2. Power source: A high voltage power source is applied across the laser machine to produce laser beam.
3. CAM: It is a computer aided manufacturing in which the laser machine is integrated with the computers to perform welding process. All the controlling action during the welding process by laser is done by CAM. It speeds up the welding process to a greater extent.
4. CAD: It is called as Computer aided Design. It is used to design the job for welding. Here computers are used to design the work piece and how the welding is performed on it.
5. Shielding Gas: A shielding gas may be used during the welding process in order to prevent the w/p from oxidation.

Types of Laser used

1. Gas lasers: It uses mixtures of gases as lasing medium to produce laser. Mixtures of gases such as nitrogen, helium and co_2 are used as lasing medium.
2. Solid-state laser: it uses several solid media such as synthetic ruby crystal (chromium in aluminum oxide), neodymium in glass (Nd: glass), and neodymium in yttrium aluminum garnet (Nd-YAG, most commonly used).
3. Fiber laser: The lasing medium in this type of laser is optical fiber itself.

Characteristics of Laser Beam Welding

1. The power density of laser beam welding is high. It is of the order 1 MW/cm^2. Because of this high energy density, it has small heat-affected zones. The rate of heating and cooling is high.
2. The laser beams produced are coherent (having same phase) and monochromatic (i.e. having same wavelength).
3. It is used to weld smaller sizes spot but the spot sizes can vary from 2 mm to 13 mm.
4. The depth of penetration of the LBW depends upon the amount of power supply and location of the focal point. It is proportional the amount of power supply. When the focal point is kept slightly below the surface of the work piece, the depth of penetration is maximized.

5. Pulsed or continuous laser beams are used for welding. Thin materials are weld by using millisecond-pulses and continuous laser beams are used for deep welds.
6. It is versatile process because it is capable of welding carbon steels, stainless steel, HSLA Steels, aluminum and titanium. Due to high cooling rate, the problem of cracking is there when welding high-carbon steels.
7. It produces high quality weld.
8. This welding process is most popular in automotive industry.

Working

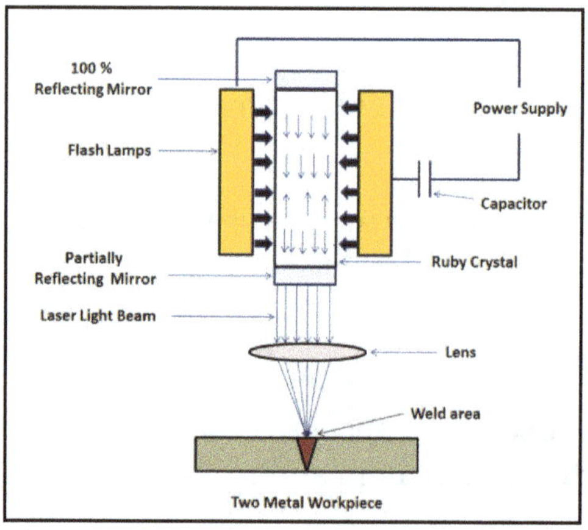

- First the setup of welding machine at the desired location (in between the two metal pieces to be joined) is done.
- After setup, a high voltage power supply is applied on the laser machine. This starts the flash lamps of the machine and it emits light photons. The energy of the light photon is absorbed by the atoms of ruby crystal and electrons get excited to their higher energy level. When they return back to their ground state (lower Energy state) they emit a photon of light. This light photon again stimulates the excited electrons of the atom and produces two photons. This process keeps continue and we get a concentrated laser beam.
- This high concentrated laser beam is focused to the desired location for the welding of the multiple pieces together. Lens are used to focus the laser to the area where welding is needed. CAM is used to control the motion of the laser and work piece table during the welding process.
- As the laser beam strikes the cavity between the two metal pieces to be joined, it melts the base metal from both the pieces and fuses them together. After solidification we get a strong weld.
- This is how a laser beam welding works.

Advantages

- It produces high weld quality.
- LBW can be easily automated with robotic machinery for large volume production.
- No electrode is required.
- No tool wears because it is a non-contact process.
- The time taken for welding thick section is reduced.
- It is capable of welding in those areas which is not easily accessible.
- It has the ability to weld metals with dissimilar physical properties.
- It can be weld through air and no vacuum is required.
- X – Ray shielding is not required as it does not produce any X-Rays.
- It can be focused on small areas for welding. This is because of its narrower beam of high energy.
- Wide variety of materials can be welded by using laser beam welding.
- It produces weld of aspect ratio (i.e. depth to width ratio) of 10:1.

Disadvantages

- Initial cost is high. The equipment used in LBW has high cost.
- High maintenance cost.
- Due to rapid rate of cooling, cracks may be produced in some metals.
- High skilled labour is required to operate LBW.
- The welding thickness is limited to 19 mm.
- The energy conversion efficiency in LBW is very low. It is usually below 10 %.

Application

The laser beam welding is dominant in automotive industry. It is used in the area where large volume production is required.

Hybrid Laser Welding

Hybrid laser welding is a process, where laser welding and some other welding process, such as MAG welding, are combined to a single process. Laser beams and electric arcs are quite different welding heat sources, but both work in a gaseous shielding atmosphere at ambient pressure that makes it possible to combine these heat sources to a unique welding technique referred to as hybrid laser-arc welding. Hybrid laser-arc processes, especially hybrid laser-MAG welding, have been increasingly applied to welding applications during recent years. In hybrid laser-arc welding process most of the process mechanisms of autogenous laser welding and arc welding can be observed. Furthermore some synergistic effects originating from the combination of two different processes are present.

Hybrid laser-arc welding is a combination of laser welding and arc welding. The processes create a common weld pool which moves along the weld pass. In some applications, the distance between laser beam and arc is so long, that the processes are actually creating separate weld pools. This process is sometimes referred as tandem laser-arc welding rather than hybrid laser-arc welding. Hybrid process is not just a simple combination of laser beam and arc, but there are complicated physical interactions between the two heat sources.

Development of Laser-arc Hybrid Welding

In recent years, laser welding has begun to be used for assembly welding of automotive bodies and parts, although it has not yet to be used widely. One restraint is that in laser lap welding, the gap between the lapped sheets must be controlled very tightly. If the gap is wide, burn-through occurs, and, if the gap is excessive, the two sheets cannot be welded together. For this reason, the gap is generally restricted to 0.1 mm or less for laser welding. Laser lap welding of zinc-coated sheets poses additional problems. Laser lap welding of zinc-coated sheets is performed without any gap, so the zinc that is evaporated between the sheets tends to blow off weld metal, or the zinc vapor tends to remain in the weld metal and form blowholes.

Laser-arc hybrid welding was developed to solve these problems. This method combines YAG laser welding and arc welding, allows a larger gap between lapped sheets than in laser welding, and produces fewer blowholes, even in lap welding of zinc-coated sheets. Therefore, stringent gap control is not necessary for lap welds, and the industrial application of this method is easy. In addition, the welding speed can be equivalent to that of laser welding, so the high efficiency of laser welding can be utilized.

Configuration of Laser-arc Hybrid Welding

Fig.1 shows the system configuration for laser-arc hybrid welding. A YAG laser is used for laser welding, and an arc-welding electrode is positioned behind the YAG laser radiation point. The aim position of the arc is about 1to 3 mm behind the laser radiation point. A YAG laser is used because the plasma does not absorb much of the laser energy. Most of the energy reaches the sheets, so that the YAG laser energy is efficiently utilized for welding. In contrast, energy from a carbon-dioxide laser is strongly absorbed by the arc plasma, so that a sufficient distance must separate the arc and the laser radiation point. Therefore, the combined effect of the laser and arc is not possible with a carbon-dioxide laser.

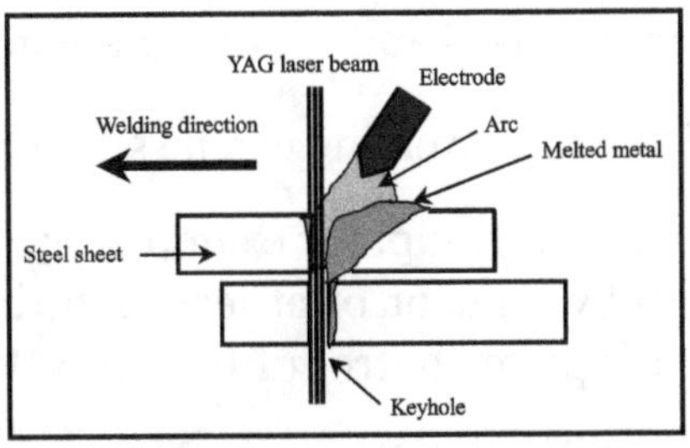

Fig: Configuration of laser beam and arc-welding electrode in laser-arc hybrid welding.

In hybrid laser-arc welding processes the laser beam as a high-energy density heat source commonly serves as the primary heat source enabling deep penetration mode welding while the arc as the secondary heat source improves process stability, reliability and efficiency as well as the quality of the weld. As a secondary heat source, either process with consumable electrodes, i.e. MAG process, or process with non-consumable electrodes, i.e. TIG process, can be chosen. If any addition of filler material is required, a processes with consumable electrode should be applied, otherwise arcs with non-consumable electrodes are preferred. Depending on the energy ratio of the two heat sources, the characteristics of the hybrid process may be either more arc-like or more laser-like. The weld penetration increases with an increase in laser energy, whereas the weld width increases with the arc energy. For different applications, there exists an optimum ratio between the powers of the laser beam and arc.

Advantages

Hybrid welding is already being used, or under consideration, by the following industries:

- Road transport – the high welding speed of the hybrid process is attractive to the high production volume environments found in the automotive industry, especially given the greater part fit-up tolerance than that of autogenous laser welding.
- Shipbuilding – the lower heat input and distortion introduced when hybrid welding, when compared with MAG welding or SAW, reduces the costs associated with distortion correction and rework. By contrast, conventional arc welding methods and their associated re-work have been estimated to constitute up to 20-30% of overall manufacturing costs.
- Rail transport – as in shipbuilding, the low distortion that can result from hybrid railcar seam welding, compared with conventional arc welding processes, is of interest as a means of reducing fabrication costs, as well as the higher welding speeds reducing overall welding time.

- Oil and gas – the hybrid welding of pipes has been well demonstrated, and with continuing developments in laser sources and pipe steels, continues to be of interest as a future means of increasing overall joint completion rates, depending on steel grades used and operating environment requirements.

The chief benefits of hybrid laser-arc welding can be summarized as:

- Improved tolerance to joint fit-up: for example, hybrid welding can extend the tolerance to joint gap by a factor of at least 2-3 times over laser welding, or greater, if real-time adaptive control of hybrid welding parameters is carried out.
- Improved weld quality: hot cracking (e.g. in some higher strength Al alloys) can be avoided, and internal porosity content reduced, with respect to laser welds.
- Increases in single pass penetration depth: this is controlled principally by the choice of laser and welding parameters used, but single pass penetrations >6-12mm can be achieved using higher power (≥5kW) lasers.
- Increases in welding speed: this is also dependent on the laser used and materials being welded, but speeds of >5m/min can be possible in thinner materials.
- The increases in penetration depth and/or welding speed, when compared with arc welding, are particularly significant: the net heat input can be reduced, resulting in lower distortion, rendering hybrid welding of particular interest for the making of long seam welds between plates or between sections, the welding on of attachments etc.

Laser Shock Peening

LSP is a mechanical (cold working) process where pulses hit the surface with high power intensity and shock waves are generated. These waves plastically deform the surface and compressive stresses are extended into the subsurface. These dynamic compressive stresses are highest on the surface and decrease with depth into the material.

Generation of Laser Shock Waves

The basic steps for laser shock wave generation can be described as the following sequence; (a) the target surface is covered with an absorbent (sacrificial) coating. This layer vaporizes, forming plasma on the surface with short duration pulse pressure. Absorbent layer prevents melting and laser ablation while maintaining high surface quality and without this layer, the energy conversion from pressure to shock cannot be made effectively. This layer can be aluminum, copper, lead, vinyl tape, zinc or black paint. Hong *et al.* showed experimentally that black paint layer has the best absorption ability to laser; where almost 100% laser energy is absorbed by black paint as compared to 80% absorption by Al. This means that when a target is irradiated, almost 100% of the laser intensity is used to generate the plasma; (b) transparent overlay (tamping layer) is applied to prevent the plasma from expanding away off the surface thereby increasing the intensity of the shock wave. These overlays, also known as the confining medium, can be water, quartz or glass. Other confining media are K9 glass, Pb glass, perspex or silicon rubber. The choice of confining medium solely depends on the substrate material, density and acoustic velocity which gives the so-called "acoustic impedance effect". Finally, plastic deformation occurs when the shock wave stress exceeds the strength (dynamic yield strength, σdyny) of the metal. This LSP process is schematically shown in figure below. The plastic deformation results in high dislocation multiplication and movements which affect the microstructure and properties of the material. Peyre *et al.* showed that increasing laser shock pressure increases the depth of plastic deformation. The description of how shock waves are generated is simple compared to the complexity of the process parameters and optimization.

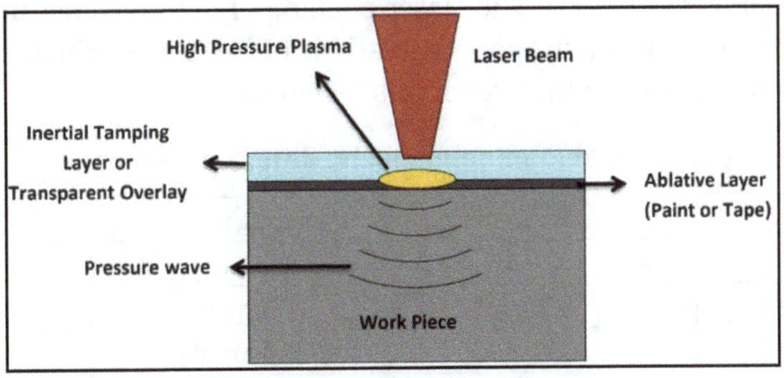

Figure: Laser shock peening (LSP) process.

Process Parameters for Laser Shock Peening

Effective shock peening process depends on the material (target), laser and beam parameters as well as the absorbent and transparent overlays properties. Typical requirements are: Q-switched laser system based on neodymium such as Nd-doped glass (Nd:Glass), yttrium aluminum garnet (Nd:YAG), yttrium ortho vanadate ($Nd:YVO_4$) or ytterbium-doped YAG (Yb:YAG) lasers. Different laser wavelengths of 1054 nm (infra-red), 532 nm (green), and 355 nm (ultraviolet) are used. Additionally, time duration of shock wave pressure (laser pulse length) ranges from 10 to 100 ns with energy of 1–100 J per shot, and spot (diameter) size of 1–6 mm. For the laser spot geometry; circular, elliptical, rectangular and square geometries have been studied with square showing better uniformity in intensity and good overlapping rate. A frequency of less than 1 kHz and power density (I) in the range of 0.1–10^6 GW/cm^2 are also required. Comparing with UIP, a higher frequency of 15–100 kHz is employed. Furthermore, the most important LSP process parameters and equations have been summarized. Power density is expressed as a function of frequency, pulse time, power and spot area as given in first equation. Other parameters are the laser fluence as given in second equation, the load or pressure (P), which must exceed the dynamic yield strength for plastic deformation given by different authors in third to ninth equations. Tenth equation shows the reduced shock impedance (Z) for the target material (Z_1) and confining medium (Z_2) which is related to the density and speed of sound in the material. Table below lists typical values of acoustic shock impedance for various confining media and target materials. Furthermore, Eleventh equation shows the Hugoniot Elastic Limit (*HEL*), which is the ultimate stress a material can withstand under compression without internal rearrangement. The process optimization could be more complicated especially when many parameters are involved. For instance, shock pressure representation was inconsistent as found in the literature. Third to ninth equations indicate how the constants were varied based on the experimental observations. The variation of constants were attributed to the type of coating, confining medium and target material properties. This suggests that selecting the most effective processing parameters is paramount. The parameter selection can be achieved by extensive experimental work and simulations. Attempt has been made by *Society of Automotive Engineers* (SAE) to issue recommended parameters for laser peening of metals. Most importantly, successful laser peening experiments and simulations have been conducted using different processing parameters as summarized in table below.

Table: Typical values of Z_1 and Z_2 as reported by different authors.

Target material	$Z_1 \times 10^6$ (g/cm²s)	Confining medium	$Z_2 \times 10^6$ (g/cm²s)
Ti-6Al-4V	2.75	Water	0.17
AA7050-T7451	1.50	Perpex	0.32
SS304	3.61	Silicon rubber	0.47
Mg-Ca	0.88	K9 Glass	1.14, 1.5
AISI 4140	3.96	Quartz Glass	1.31
Cu	0.16	Pb Glass	1.54
SS321	4.00	BK Glass	1.44

Table: Typical LSP processing parameters from the literature used for different materials with concluding remarks.

Material	Power density (GW/cm²)	Pulse duration (ns)	Absorbent coating	Transparent overlay	Remarks
Ti-6Al-4V	5	10	Al Foil	Water	The fatigue life was improved by 22.2% and 41.7% for one and two successive laser shocks respectively as compared to as-received sample.
Ti-17	3, 6, 9	9, 27	Al	Water	It was found that all LSP parameters had influence on the residual stress profile. LSP had no influence on roughness, little effect on work hardening and large effect on hardness due to the compressive stresses induced.

AA2195	5	18	Al	Water	The fatigue crack growth rate of stir welded joint was reduced using LSP as compared to SP. Furthermore, the reduced crack growth rate was comparable with unwelded material.
Al2024 T3	5	18	Black paint	Water	LSP specimen showed resistance to fatigue crack growth for various notch geometries as compared to unpeened specimen.
4140 Steel	3	5	Al	BK7 Glass	WLSP and DSA showed improved fatigue life due to enhanced cyclic stability of compressive RS. The stability and reliability were attributed to the enhanced dislocation pinning effect associated with the WLSP and DSA process.
316 L SS	2.5	10	No coating	Water	The potentiodynamic polarization investigation showed improvement in corrosion potential (Ecorr) and corrosion current density (Icorr) with increase in laser pulse density. LPwC specimen showed 30%–40% increase in microhardness than unpeened specimen.
2204 Duplex SS	900, 1600, 2500 pul/cm^2	8	No	Water	Greater RS were observed with higher pulse density. LSP showed improved crack growth rate resistance and fracture toughness as compared to untreated specimen.
Cu 15 µm foil	6	10	Black paint	Water	The microscale laser dynamic forming (µLDF) showed that, materials are strengthened due to refined grains and large dislocation density which were dominant in the microstructure.
Alloy 22	10	25	Al	Water	Slitting method was used to establish a relationship between LSP parameters and RS on flat coupons. While contour method measured the spatial RS on critical geometrical features.
Inconel 600	10 J/cm^2	12	-	Water Jacket	Tilted column microstructure was observed due to overlapping of laser spots during scanning. Spherical nanoparticles with 60 nm diameter were seen due to laser ablation.
Fe-3%Si	0.1–1	25–200	Pb	Quartz	Laser shocking showed that quartz plus lead overlay gave the most deformation for a given power density. The deformation occurred by slip and twining mechanism. Most of the samples showed surface melting for 200 ns pulses.

Mg-Ca	78	5 to 7	Black paint	Water	Sequential peening showed an increase in the tensile pile up region up to 50% which was considered applicable for orthopaedic applications.
AZ31B Mg	5	23	Al7075	Water	Sub-grain sizes of 5.8 μm were obtained after four laser impacts. The RS near the surface increased with number of impacts. LSP showed retardation to SCC crack initiation and propagation.
Brass H62	1000, 2000, 3000 Pulses/ cm^2	10	-	Water	LSP showed that, the higher the pulse density the higher the microhardness, roughness and wear resistance. No observable features in the microstructure after LSP.
CMSX, AM1, Astroloy	12–150, 90, 25	25, 33, 40	Black paint	Water, Glass	The deformation microstructure and macroscopic mechanical phenomenon studies were envisioned to provide optimum process parameters applied to fatigue and fretting fatigue specimens.

Intensity :

$$I\,(\frac{GW}{cm^2}cm^2)=\frac{P_{avg}}{f\ (pt)a}$$

Laser Fluence:

$$Fluence\ (\frac{J}{cm^2})=\frac{Laser\ pulse\ energy\ (J)}{focal\ spot\ area\ (cm^2)}$$

Pressure:

$$P=A(\frac{\alpha}{2\alpha+3})^{1/2}Z^{1/2}I^{1/2}$$

$$P(GPa)=0.01(\frac{\alpha}{2\alpha+3})^{1/2}Z^{1/2}I^{1/2}$$

$$P(GPa)\ =1.02I^{1/2}$$

$$P(Pa)\ =0.25(IMA)^{1/2}$$

$$P=5.47(\frac{\alpha}{2\alpha+3})^{1/2}I^{1/2}$$

$$P\ (kbar)\ =BI^{1/2}$$

$$P=(AZI)^{1/2}$$

Reduced Impedance:

$$\frac{2}{Z}=\frac{1}{Z1}+\frac{1}{Z2}$$

Hugoniot Elastic Limit (*HEL*):

$$HEL=\frac{1-\theta}{1-2\theta}\sigma_y^{dyn}$$

where, I is the laser power density in GW/cm^2, P_{avg} is the average power output in W, f is the laser frequency in Hz, pt is the pulse time in ns, a is the laser spot area in cm^2, B is 21 or 10.1 for glass- or water-confined modes, respectively, A is absorption coefficient for surface coating, M is transmission coefficient for transparent overlay, α is the ratio of thermal to internal energy, Z is the reduced shock impedance of the target material and confinement medium, θ is Poisson's ratio and σdyny is dynamic yield strength at high strains.

Typical processing parameters presented in Table above are power density, pulse duration, absorbent coating and transparent overlay. However, despite the fact that the same material is laser peened, the processing parameters varied for different authors. This indicates that the choice of processing parameters is solely confined to laser type, substrate condition and target application. The characterization of LSP treated surface will be instrumental in understanding the LSP process.

Laser Peened Materials

Several materials have been peened successfully proving the effectiveness of LSP and the associated residual stresses induced. Typical research studies conducted are influence of LSP process on residual stress prolife, microstructure, mechanical properties, fatigue life, fretting fatigue life, corrosion and stress corrosion cracking behavior. Recent LSP studies have shown that laser peening technique can be used for shaping and forming of metals and inspection of coatings. The materials laser peened include titanium alloys, aluminum alloys, different steel grades, copper alloys, zinc, nickel based alloys, super alloys, brass, magnesium alloys, bulk metallic glass and other materials. Comparatively, similar materials have been treated by the UIP technique such as soft steel, 304SS, Type 35 carbon steel, 316L SS, Inconel 718, naturally exfoliated Al7075-T6511, AlSi11Mg, Al2024-T351, Al matrix composite reinforced with AlCuFe or Ti fine powders, Ti alloy, welded joints of steel grade 345W, steel grade A36, 16Mn steel and 304SS. Despite treating similar materials only a few researchers compared LSP and UIP processes.

Characterization of Laser Shock Peened Materials

Due to the high strain rates of $10^6 \cdot s^{-1}$ involved in the laser shock peening process, it is usually accompanied with changes in microstructure and mechanical properties (hardening and strengthening). Despite the high strain rates during LSP, the total induced strain is much less than in the case of shot peening. Furthermore, these microstructural changes are characterized and investigated by means of TEM and SEM for the surface and sub-surface morphologies and features. For the TEM technique, typical information revealed are dislocation structure, dislocation entanglement, dislocation band and precipitate structure. SEM technique on the other hand, provides surface and subsurface features such as surface micro cracks, macro cracks, fretting cracks, mild indentations, melting due to ablation and other surface irregularities from the LSP process. The measured compressive residual stress induced during the LSP is an important indicator of the effectiveness of the LSP process. Since residual stresses cannot be measured directly due to its extrinsic nature, strain in the crystal lattice which is an intrinsic property is usually measured. The residual stress are calculated from the strain while assuming a linear elastic distortion of the crystal lattice. Several methods have been used to quantify the residual stress profile in the material. Among these methods are X-ray diffraction $\sin^2\psi$, hole drilling, slitting, neutron diffraction and contour methods. X-ray diffraction $\sin^2\psi$ method is the most commonly used method to quantify the residual stresses. In this method, the strains are measured at particular diffraction angles and lattice planes where Bragg's law is satisfied. Material constants such as Young's modulus, Poison's ratio as well as different tilt angles "ψ" are further employed to quantify the stress. For in-depth or subsurface residual stress profiles, measurements after material removal by electro polishing is used extensively. The step-by-step material removal is in the order of microns and more details of this procedure have been reported in. For the hole drilling method, a hole is drilled and variations of stresses are measured as a function of depth by a strain gage rosette. Surface roughness measurement after LSP has been reported as the arithmetic average (Ra) using surface profile meters such as Talysurf contacting surface and electronic contact (stylus). The roughness can be areas of stress concentration and crack initiation sites which have adverse effect on fatigue life of components. Micro hardness and nano hardness across the irradiated surface are also measured and reported in the literature. Increase in measured hardness has been attributed to plastic deformation resulting in dislocation multiplication and movement. For in-depth or subsurface hardness measurements, nano hardness is a more reliable approach as compared to micro hardness. This is because of the compressed layer thickness which might pose measurement difficulty when measuring micro hardness. However, these characterization techniques give researchers an insight regarding the changes and modifications due to LSP processing. The general improvements can be further quantified after a series of mechanical testing depending on the application, such as test for fatigue resistance, fretting fatigue resistance, stress corrosion cracking, and water erosion resistance.

Advantages of Laser Shock Peening

One advantage is that in laser peening, the depth of compression exceeding shot peening produced with low cold work. The subsurface compressive residual stress distribution tends to be maximum at the surface, diminishing nearly linearly with depth. The low cold work provides thermal and mechanical stability. an advantage in hot section engine applications, or where there may be momentary overload due to impact, etc.

Disadvantages to Laser Shock Peening

Disadvantages associated with laser peening include:

- High capital cost of the laser peening equipment, which may include a clean room environment and million dollar plus installations.
- Laser shock peening requires repeated coating with tape to produce the depth and magnitude of compression achievable in a single pass.
- Because the explosive laser shocks remove the tape from adjacent treatment sites, two coating cycles are usually required to achieve one layer of treatment. Multiple layers are generally required to achieve the desired depth of compression.
- Process quality control has slowed the application of laser shock peening due to the difficulty of controlling the processing variables involved. Variation in the thickness and even turbulence of the water layer tamp, as well as debris in the air and on the mirrors affect the compression achieved, even when the laser power is precisely controlled. The laser shock peening process cannot provide the closed-loop control available with LPB. Internal cracking caused by the superposition of shock waves from opposing faces of blades, or echoing waves from the opposite wall of thicker sections has been found to limit the application. The cost of laser shock peening is the highest of any of the surface enhancement methods, typically 10 to 100 times shot peening.

Laser Engraving

Laser engraving is the process of using a laser to etch or mark the surface of a work piece. Compared to other engraving methods, laser engraving is often considered to have many advantages, though it also has some disadvantages. For instance, it can be extremely precise, but equipment costs are often prohibitive. This technology can be useful for many applications and can utilize many materials. Engraving methods vary, depending on the type of work piece and the material, which means that different engraving equipment is sometimes required.

Uses for laser engraving are usually based on the production of a specific design. The type of design often dictates the use. Professional engraving can include industrial or commercial production of engraved products. These can include duplicate images or customized images at nearly any image quality, though high-quality engraving is often more costly than low-quality engraving. Partly because of the wide range of materials that can be used, engraving also can be used for artistic work.

Stone, glass, metal, plastic and natural materials can all benefit from laser engraving. In fact, for some of these materials — such as stone and metal — engraving can be the primary form of marking for any purpose. Harder materials such as stone and glass are often difficult to engrave using other forms of engraving. For instance, laser crystal engraving relies heavily on the use of lasers to create images inside of hard, crystalline material. The variety of materials used results in there being many different types of engraving.

Industrial engraving, for instance, might make use of large machines to mass produce a single laser-engraved product or design. Professional engraving, on the other hand, might rely on small machines to customize products based on consumer needs. Despite these differences, the hardware used in laser engraving is often similar.

The hardware used in laser engraving is often constructed in three major parts: the laser, the working space and the interface. Lasers can be constructed in many ways, and some machines might have complex lasers while others rely on simple lasers. Working space often depends on the type of laser being used. For instance, simple engraving equipment might have a flat table for the work piece, whereas complex engraving equipment might hold and move the work piece on multiple axes. The interface of these systems is usually a computer into which the specific designs are put and through which the engraving commands are carried out.

Working of Laser Engraving

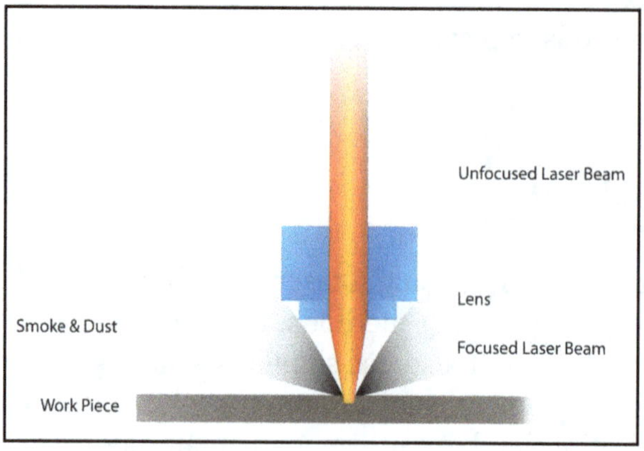

The high heat of the laser beam vaporizes the material thereby cutting into the part's surface and physically removing material. This process leaves a cavity in the surfaces that is not only visible with a high contrast but also noticeable by touch. There are noteworthy differences in the depth of this cavity that varies between 0.02″ in metals to 0.125″ in harder materials.

Usually, the engraved areas become black. Multi-layer materials offer an alternative, as they enable to engrave other colors than black. By removing the top layers the lower layers become visible.

Another variant of laser engraving is relief engraving. The laser creates a deep engraving with distinct heights. This variant is beyond any doubt the most impressive, since the overall effect is that of a wooden sculpture.

Laser Engraving on Different Material

It is possible to laser engrave finished leather, producing a look similar to hot-branding. Avoid raster marking large surfaces, since the leather will turn powdery. Therefore, vector marking is recommended.

Since leather is a natural material, it may deform during laser engraving if too much power is applied.

Because of the many varieties of leather it is difficult to come up with a general rule for laser engraving it. Hence, there is no way around determining the optimal laser engraving settings for each kind separately. To this end test your machine with a gray scale matrix

Laser Engraving Paper/Cardboard

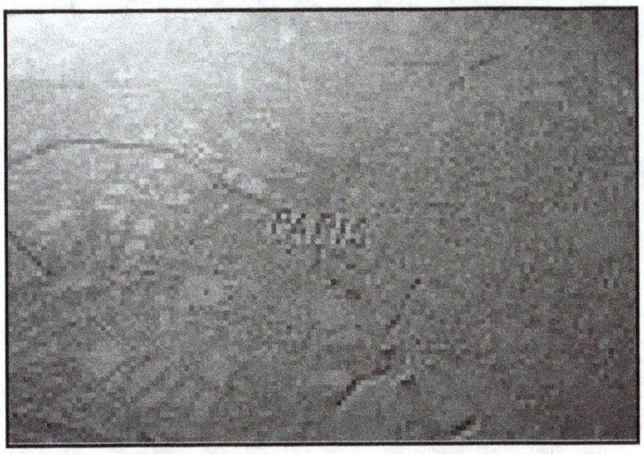

Because of the many varieties of paper and cardboard it is advisable to determine the optimal laser engraving settings for each kind separately. To this end test your machine with a gray scale matrix.

Laser Engraving Wood

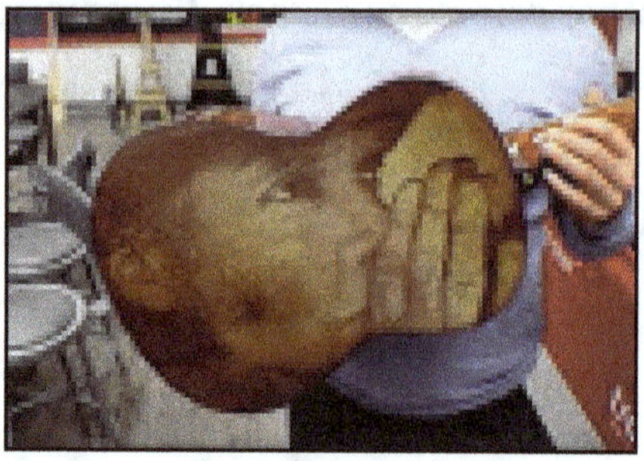

Besides natural woods there is also plywood, which has several advantages. This laser engraving material consists of several sheets of wood that are glued together. As a result, plywood boasts a raised mechanical resistance, water resistance and lightness. All these properties have made plywood a staple for builders, designers, artists, and craftsmen.

In terms of laser engraving, plywood has an additional advantage over natural wood. When cut into sheets, the latter reveals grains in alternating shades that stem from different kinds of growth during the seasons. These differences also translate into different laser engraved surfaces. Usually, the lighter areas will come out lighter and the dark areas will be darker.

As a rule of thumb, unless the desired look of your design is kind of zebra pattern, you will be better off using plywood that has a uniform surface.

In addition, you should always apply higher laser power when engraving hard wood. Only then will you smooth results.

To determine the optimal laser engraving settings, test your machine with a grayscale matrix.

Laser Engraving Glass

The effects the laser has onto the glass surface can be manipulated using a number of practical methods.

Apply a moist paper towel to achieve a smoother surface finish and a white engraving result. Take care that there are no air bubbles or overlaps, because this would distort the effect. Once the engraving is finished the paper towel can be wiped off easily.

An alternative to paper towels is application tape. Like moist paper towels, application tape is an excellent measure against surface roughening. The same caveat concerning air bubbles and overlaps apply. However, the final engraving color is not a bright white, but looks grey. Once the engraving is finished the tape residue can be removed off easily.

But what to do when your laser engraver has not enough power to engrave glass? Even then you can achieve designs with high-contrasts! Simply mask the glass surface with tape. Then you engrave the design into the tape. Next, cover the revealed area of the tape with a thin layer of paint using a brush. After the paint has dried, peel the tape and remove the tape residue carefully.

Laser Engraving Stone

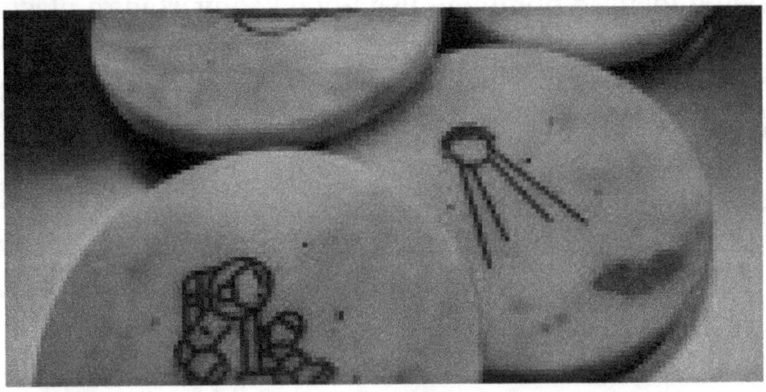

Experience shows that it can be piece of work to laser engrave stone. You will get the best results with polished stones. To avoid distortions of the laser engraving motif, try to use stones that have a flat surface. You can align natural stones with the laser by using modeling clay as a support. When you set the laser's focus, apply an average value to achieve a uniform result.

If your results are poor, consider changing the Z-offset value. This simple trick can significantly improve the result of your engraving. Many professional laser engravers report that hard stones are easier to process if you choose a negative Z-offset value of 1mm or even more. The physics behind this workaround are simple. The closer the surface to be engraved is to the laser, the more energy enters.

But what to do when your laser engraver has not enough power to engrave stone? Or, when working with extremely hard stones that yield no satisfying results? Even then you can achieve designs with high-contrasts! Simply mask the glass surface with tape. Then you engrave the design into the tape. Next, cover the revealed area of the tape with a thin layer of paint using a brush. After the paint has dried, peel the tape and remove the tape residue carefully from the stone's surface.

Laser Ablation

Laser ablation is the process by which layers are removed from solid metals and industrial compounds using a laser beam for ultimate precision. The beam will irradiate the surface, meaning that it has been exposed to radiation.

Laser Ablation Process

Laser ablation works by focusing a laser onto a substrate to a remove material that is on its surface. The amount that is removed depends on the intensity, pulse length, and wavelength of the laser, as well as the material itself. The area absorbs the laser that is being directed on it, thereby breaking down the chemical bonds within the area.

With a low level of laser flux, the material being focused upon is absorbed by the laser's energy and then changes to a gaseous state. With a higher level of laser flux, the material that is being focused upon will usually be converted to plasma. It is possible to undergo the laser ablation process with both a pulsed fiber laser and a continuous wave laser, although the former is the more common method due to the high level of laser intensity.

Laser ablation has many benefits over more traditional methods of processes, such as with thin film removal, whereby alternative solutions have to undergo a multi-step process which is costly, time-consuming and inflexible, as well as having risks for the environment. Laser ablation is a much more efficient, reliable and cost-effective method.

Uses of Laser Ablation

Laser ablation has many different uses, particularly as the whole process is undergone with minimal excess heat being transferred to the surrounding area of the material being used. This means that the process will have little to no effect on the parts of the material that you don't want it to, and so laser ablation is great for a wide range of materials, including plastic, metal, ceramic, and even biological tissue. With the advanced nature of fiber lasers, you can focus on a very small, specific spot to produce your desired results.

Laser ablation can also be used to determine the presence and concentration levels of a particular chemical or material on a surface. This is achieved by generating bright plasma on the surface, and then analysing this plasma to see what is present there. This is a much more environmentally-friendly process for determining chemical analysis, as opposed to more traditional methods such as using toxic acid solutions.

Finally, laser ablation can be used to deposit film on a surface, so as to produce things like Nanomaterials or to fabricate superconducting materials. This can also be used for the micro-structuring of an object, whereby the structuring of a larger object can be seen.

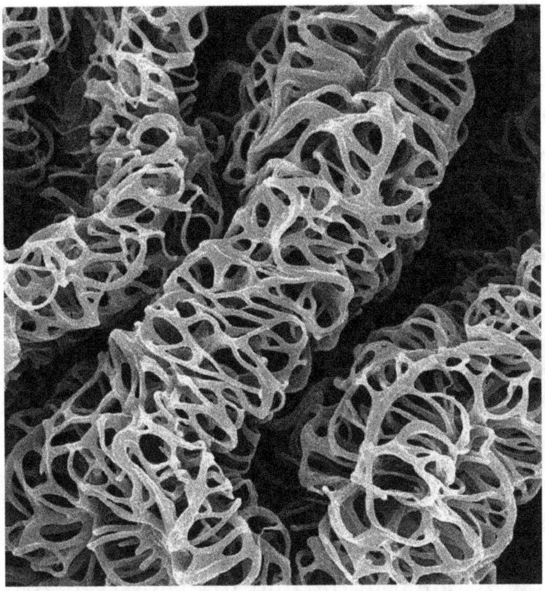

Fig: Laser ablation is also useful for viewing the micro-structuring of an object.

- Factors Need to be Considered for Laser ablation

The success and efficiency of the laser ablation process depend on a number of factors. These are:

- Wavelength

The wavelength needs to be selected carefully with a minimum absorption depth. This is so that there is a high energy deposition within a small volume, resulting in a quick and efficient laser ablation.

- Pulse duration

It is better to use shorter pulse durations so as to perform at maximum peak power and to minimise the thermal damage that is inflicted on the surrounding area, although this is already greatly reduced due to the precise nature of the laser ablation process.

- Pulse repetition rate

The pulse repetition rate needs to be high enough so that the heat caused by laser ablation is maintained and is not left any time to cool. This will result in a more successful ablation as energy waste will be reduced.

- Beam quality

Your laser beam needs to be of sufficient quality to perform the laser ablation process successfully. The quality of the beam will be determined by its brightness, its focus ability, and its homogeneity. The beam size will also need to be controlled so that you don't ablate too large an area.

- Application Areas for Laser Ablation

Due to the precise, accurate and environmentally-friendly nature of laser ablation, it is used in many different application areas and industries. Its ability to remove thin film has particular benefits for the electronics and semiconductor industries, and it has also found uses for other industries such as medical, automotive and ship-building. It has understandably become the preferred solution for many manufacturing processes.

Laser Bonding Technology

Laser bonding is an established technology that uses lasers to bond additive marking material to a wide variety of different substrates and is commonly used with CO_2, Nd: YAG and fiber lasers.

The technology was first invented in the mid 1990s by Paul W. Harrison, the founder of The Mark. After selling his interest in The Mark, he went on to found Laser Bonding Technology and continues to be involved with laser marking hardware and the further development of new and unique marking materials. These new products produce permanent marks on most metal, glass, ceramic and stone surfaces for use in a wide range of industrial and artistic applications, ranging from automotive, aerospace and medical to the awards & personalization industries. It differs from the more widely known techniques of laser engraving and laser ablation in that it is an additive process, adding material to the substrate surface to form a permanently bonded mark or decoration instead of removing material as is done in those other laser processes.

For metals, parts can be permanently marked with high contrast and high resolution for identification, logos, bar-coding and serialization purposes, without damage to the substrate. With glass and ceramics, complex surfaces can be decorated or marked; however, the traditional kiln firing process is replaced by a laser, with its permanently bonded decorations and markings fired on in seconds instead of minutes or hours.

References

- Laser-shocking: lambdatechs.com, Retrieved 19,June 2020
- What-is-laser-engraving: wisegeek.com, Retrieved 10,July 2020
- Hybrid-laser-welding, laser-processing, technologies: ionix.fi, Retrieved 29,April 2020
- Electrical-discharge-machining-EDM-424360: britannica.com, Retrieved 11,May 2020
- Laser-beam-welding: theweldingmaster.com, Retrieved 21,March 2020

4

Laser Cutting: An Integrated Study

The technology which makes use of lasers in order to vaporize materials for the purpose of producing a cut edge is known as laser cutting technology. A few different methods of laser cutting are vaporization cutting, stealth dicing and thermal stress cracking. The topics elaborated in this chapter will help in gaining a better perspective about laser cutting as well as these types of laser cutting methods.

Laser cutting is a way to cut precise patterns in metal, plastic, wood, and practically every other material that man works with. It allows a level of accuracy and complexity impossible with conventional machining tools. Laser cutting works by exciting a gaseous medium, commonly carbon dioxide, causing it to amplify light reflected back and forth multiple times within the laser chamber. The light emerges from an aperture and is focused by a lens onto a specific point.

A typical process laser has a beam about a 1/5 of a millimeter in width, focusing 1000 to 2000 watts of energy. This is enough to melt most common materials. Because lasers become less focused and lose energy as they penetrate through a material, there is a limit of about 20 mm for the deepness of the cut. Laser cutting machines are integrated into a larger CAD/CAM (computer-aided-design, computer-aided-manufacturing) system that takes a design file and implements it on a work piece. These machines represent a stepping stone in the continuing trend away from hands-on manufacturing, putting human workers in a more removed, creative design role.

Because a laser is made up of photons, parts of its energy can be reflected away by materials such as aluminum and copper alloys. These materials are also thermal conductors, meaning they distribute incoming heat more evenly throughout their volume. For this reason, carbon alloy and stainless steel are popular work piece materials for laser cutting. They are poor at absorbing heat, so heat is concentrated into the laser's path more readily.

Typical laser cutters allow features as small as 1 mm in size. Specialized lasers often feature sizes that are even smaller, allowing the fabrication of true micro machinery.

The first laser cuts were made in 1967, as part of a spin off of a military research project focused on lasers. Because the beams used in cutting are "class 4" lasers, the machines are designed to ensure that human operators are never exposed to them directly. All the cutting is done inside the machine.

Types of Laser Cutting

There are three main types of laser cutting:

- Co_2 laser cutting

Co_2 laser cutting is done by using a gas laser. The gas in question is a carbon dioxide mixture which has been electrically stimulated. Co_2 laser cutting is most often used on non-metal materials as they have a wavelength of 10.6 micro metres.

- Crystal laser cutting

Crystal laser cutting uses nd:YAG (neodymium-doped yttrium aluminium garnet) and nd:YVO (neodymium-doped yttrium ortho-vanadate) lasers. These are crystals which belong in the solid state group of lasers, and have the same wavelength as fiber lasers, around 1.064 micrometres. This means that they are useful for both metals and non-metals.

However, this type of laser machine is made up from expensive pump diodes, which can be costly when needed to be replaced; typically, after 8,000 – 15,000 hours of use. The crystal used also has a much shorter service life than that of a fiber laser.

- Fiber laser cutting

This type of laser cutting is completed using a fiber laser, the type of lasers that we have here at SPI Lasers. Like the method of crystal laser cutting above, fiber lasers belong to the solid state group too.

A beam is created using a 'seed laser', which is then amplified using glass fibers and pump diodes. They have the same wavelength as above, 1.064 micrometres. This means that the laser beam for a fiber laser is roughly 100 times greater than that of a Co_2 laser when it comes to intensity. It also means that they can work with both metal and non-metal materials.

A fiber laser is the most useful of the three laser types. Alongside the benefits listed above, they are often maintenance free, require much cheaper replacement parts, and have a much longer service life of around 25,000 usage hours.

Materials that can be Cut by Laser Machine

A laser cutting machine is great at cutting a huge variety of materials. Listed below are just some of the many materials:

- Wood
- Gemstones such as diamonds
- Titanium

- Steel
- Reflective metals
- Glass
- Plastic
- Silicon.

Whatever the material is that you need to work with and whatever the application, it's highly likely that laser cutting will more than be up for the task.

Characteristics of Laser Cutting

Laser cutting is most commonly used laser beam machining (LBM) process due to its different advantages such as narrow kerf, small HAZ, high accuracy and cut edge quality, material versatility, high flexibility, no tool wear, non-contact nature, high production rates and better material utilization. Due to excellent cut quality and narrow kerf, laser cutting process can be applied for a very precise cutting of sheet metal. The intricate and complex profiles with very narrow kerf width can be created in any type of material by the laser cutting process. The concentrated laser beam produces very little HAZ and laser cut specimens do not require further subsequent machining operations. Laser cutting can be applied to almost all categories of materials such as metals, super alloys, ceramics and composites. The laser cutting capability depends mainly on optical and thermal properties of the material rather than mechanical properties. The materials which exhibit high degree of hardness or brittleness and possess through favorable thermal and optical properties such as low reflectivity, low thermal conductivity and diffusivity are well suited for laser cutting process. As laser beam can be transferred from one place to another place with the help of optical fiber, it can be applied for cutting in inaccessible areas.

According to the different types of lasers are available in market, solid state Nd: YAG and gaseous CO_2 lasers are mostly used for cutting in industries due to their high powers and other suitable properties. The use of CO_2 lasers is not common in practice for cutting of highly reflective and thermally conductive materials such as copper, aluminum and their alloys due to its poor absorptivity. High reflectivity of these materials require high laser power when cut with CO_2 laser and there is possibility of damage to laser cavity, cavity optics or beam delivery optics due to reflected laser beams. Further, high thermal conductivity of aluminum alloys tends to produce large heat affected zone (HAZ), wider kerf and oxides on the molten materials in the cutting front. Although Nd:YAG lasers have low average power but when operate in pulsed mode, give higher peak power and beam intensity. Due to the shorter wavelength (1.06 μm), it can be better absorbed by reflective materials such as copper, aluminum and their alloys as compared to the CO_2 laser. Pulsed Nd: YAG laser cutting becomes an excellent means of cutting due to good focusing characteristics, smaller kerf width and narrow HAZ.

Most of the organic materials such as plastics, wood based products, leather and natural rubber cannot be cut by Nd: YAG laser as these materials are transparent to the Nd: YAG laser light and most of laser light passes through the material without heating it. But CO_2 laser can be used for the cutting of these materials.

Applications of Laser Cutting

The main industrial application of the lasers at the present time is in the laser cutting of different materials. The applications of laser cutting in different fields are numerous. Laser cutting is the neatest and the fastest profile cutting method. Since the setup time is very less, the jobs can be completed in only few minutes. The ease of the profile cutting has opened up several interesting novelties in design. It is used in an economical way for the production of many components. Laser cutting is widely used in different technologically advanced industries such as automobile, aerospace, nuclear, medical, chemical, civil structures and household applications. Mild steel sheets are most commonly used for civil structures and house hold applications which are easily cut by laser cutting.

Laser cutting is used for the cutting of titanium and its alloys which are most widely used for different industries such as aerospace, marine, chemical, food processing and medical due to their superior performance characteristics such as high strength and stiffness at elevated temperatures, high strength to weight ratio, high corrosion resistance, fatigue resistance, and ability to withstand moderately high temperatures without creeping.

Aluminium and its alloys cannot be cut easily by conventional cutting methods due to their thermal and electrical properties so that laser cutting is applied for the cutting of this type of materials. It is widely used in the market due to its unique properties such as their lightness, high strength-to-weight ratio and good corrosion resistance. These are most widely used in aircraft, automobile, ship building and chemical processing industries. In the coated form (alclad sheet), duralumin is used for the construction of ship frames, and when alloyed with nickel, it is used for making the pistons and piston rings of aircraft and automobile engines.

Besides the metals, laser cutting is also applied for the cutting of super alloys which are commonly used for making the different parts such as aerospace parts, gas engine components, parts for nuclear power plant, ship building due to its good mechanical properties at high temperature and corrosion resistance in hostile environments. Laser cutting process is also applied for nuclear power plant to dismantle and repair the plant with the help of high power lasers.

Apart from these applications, laser cutting is applied for the cutting of difficult-to-cut materials such as ceramics and composites which have many applications in mechanical, electrical and engineering applications due to outstanding thermo-mechanical and thermo-chemical properties. The traditional methods of ceramic tiles cutting, such as diamond saw, water jet machining and ultrasonic machining, are very expensive and time consuming. Laser cutting is applied for the cutting of ceramic tiles.

Short and ultrashort pulse lasers are used for the cutting of different parts at micro and nano levels. Short pulse Nd: YAG laser is successfully used in electronic industry to cut QFN packages (semiconductor packages which are plastic encapsulated packages with copper lead frame substrate). High quality femto-second lasers are used for the micro cutting of alumina ceramics which is used for the manufacturing of high power radio frequency electronic circuits and fabrication of multichip modules, mainly due to its high electrical resistance, high dielectric strength and high thermal stability.

In the medical field, different surgeons use lasers for the cutting nickel- titanium shape memory alloys which are used for different medical applications such as wires for minimal invasive surgery, super elastic devices for cardio-vascular systems, and spinal cord.

Bulk metallic glasses (BMGs) are disordered materials that lack the periodicity of crystalline structures. BMGs exhibit many exceptional physical, chemical and mechanical properties that differ markedly from those of crystalline metals. Thus, BMGs have attracted considerable attention as structural materials. Laser cutting with ultra short pulses is employed for the cutting of BMGs.

Laser Cutting Machine

Laser cutting machine adopts the energy released on the time when laser beam irradiate metal surface. The metal is melt by laser and sinter is blow away by gas. Because laser power is highly focused, only a very little heat effects the other part of metal plate and causes a little or no deformation. Laser can cut any complex shape precisely, which needs no further processing.

Laser source is generally CO_2 laser beam with operating power of 500~5000W. The power is even lower than that of many household electric heater, and because of lenses and reflectors, laser beams are focused in a very small bit of area. Highly focused energy heat the area quickly and makes the metal plate melted.

Laser cutting machine can cut stainless steal of thickness less than 16mm; when adding oxygen in laser beam, the cutting thickness is 8~10mm but it will generate a thin oxidation film in the cut surface. The maximum thickness is 16mm which leads to larger cutting deviation on the size of components.

Since the advent of laser, numerous laser products have been developed, such as laser printer, laser cosmetic instrument, laser marker, laser cutting machine etc. Due to its late start in China, the laser technology in China is greatly behind the developed countries. Although Chinese manufacturers can produce plenty of laser products, some key parts such as laser tube, driving motor, galvano meter and focus lens are imported products. This leads to an increase on cost thus an increase on consumer's payment.

In recent years, domestic research and production of laser products become closer to advanced overseas products with the progress of laser technology in China. Some aspects are even superior to products abroad, which has a leading role in market because of the advantages of price. Overseas products have absolute predominance in precision machining for its quality on stability and endurance.

Work Theory of Laser Cutting Machine

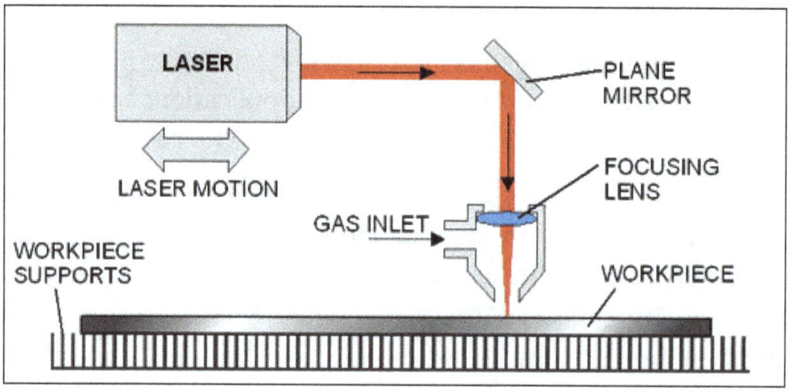

Laser tube is the core part of laser cutting machine. CO_2 laser tube.

Laser tube is composed of hard glasses, so it is fragile. It adopts layer of sleeve construction with discharge tube in the most inside layer. However, the diameter of discharge tube is thicker than laser tube, diffraction between the thickness of discharge tube and the size of flare is in direct ratio; the length of tube is in proportion to output power of discharge tube. Laser tube generates a large quantity of heat in the operation of laser cutting machine, which influences the normal work. So cold water machine is needed to cool laser tube, ensuring constant temperature for successful running.

Cutting Features of Laser Cutting Machine

- One — high efficiency

Laser cutting machine is always connected to several numerically-controlled rotary tables to achieve numerical controlled cutting. It only needs to change the NC program to adjust to components of different shapes, which can make 2D cutting as well as 3D cutting.

- Two — high speed

When cutting low carbon steel sheets of 2mm thickness, the speed of 1200W laser cutting is 600cmmin; when it is 5mm thick polypropylene resin plate, the cutting speed is 1200cmmin. The material needs no clamping fix in laser cutting process.

- Three — high quality cutting

Laser cutting features thin kerf. The two sides of kerf are parallel and the kerf is vertical to the surface. The cutting precision can reach to ±0.05mm. The cutting surface is clean and nice, with roughness of tens of microns. The cut components can even come into use directly without further machining. After laser cutting, the heat effected area is very small and material near to kerf has not been affected, making little deformation, high cutting precicion and perfect geometrical shape.

- Four — non-contact cutting

Laser cutting is non-contact cutting, which means no tool wear problem. When processing different shapes, there is no need to change tools, the only way is to alter the output parameter of laser. The whole laser cutting process features low noise, little vibration and little pollution.

- Five — various cutting material

Compared to oxyacetylene cutting and plasma cutting, laser cutting can be applied on more materials, including metal and non-metal, metal matrix and non-metallic matrix composite, leather, wood as well as fibers.

Laser Vaporization Cutting

Vaporization is the main way to remove the processed material. In the process of vaporization cutting, work piece surface is heated to vaporization temperature quickly by focused laser beams, forming high pressure steam and spraying outward at supersonic speeds. In the meantime, a hole is formed in the laser active area and laser beams reflex several times in the hole to increase the absorption of laser by material.

When high-pressure vapors spray outward, the melted materials are blown away in the kerf till the work piece is finally cut. Vaporization cutting needs very high power density, which is eighth power of ten watt above per square centimeter. It is usually applied in low flash point materials and refractory materials.

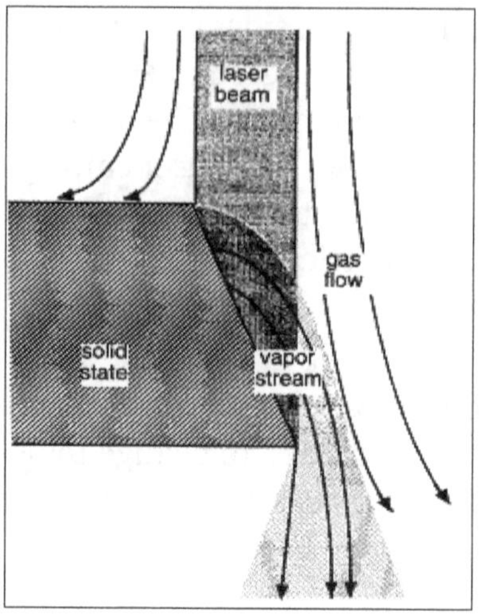

Figure: A sketch of laser vaporization cutting.

Typical materials that are cut by the vaporization method are acrylic, polymers, wood, paper, leather and some ceramics. This method has a high power requirement that depends on the thermal properties of the material. High power densities are obtained by appropriate adjustment of the laser radiation and focusing. For cutting of metals, laser vaporization cutting is the method with the lowest speed among other methods; however, it is suitable for very precise, complex cut geometries in thin work pieces.

Vaporization Cutting Formulas

Recall the velocity of melt front formulas,

$$V_s = \frac{H}{\rho(CT_v + L_v)}$$

Where, H is power density absorbed per square area and the temperature at the surface from the uniform illumination formulas for vaporization point,

$$T(0,t) = \frac{2H}{k}\sqrt{\frac{\alpha t}{\pi}}$$

Thus the time for vaporization is,

$$t_v = \frac{\pi}{\alpha}\left[\frac{T_v k}{2H}\right]^2$$

If we had a 2 KW laser focused to 0.2 mm Then average power is,

$$H = \frac{2000}{\pi r^2} = 6.3\times10^{10}\,Wm^{-2}$$

Laser Fusion Cutting

The laser fusion cutting process, also called inert gas melt shearing, is based on transformation of the material along the kerf into the molten state by heating with laser energy and the molten material blown out of the kerf by a high-pressure inert gas jet. The laser beam is the only heat source during this cutting process and the high-pressure inert gas jet is responsible for melt ejection. The inert gas jet (mainly nitrogen or argon) is also responsible for shielding the heated material from the surrounding air as well as protecting the laser optics.

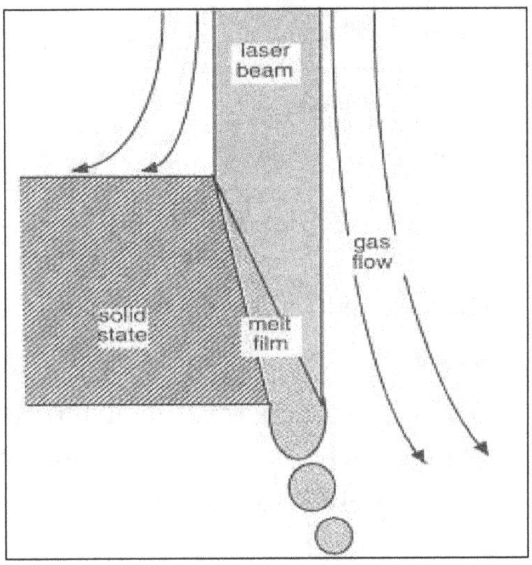

Figure: A sketch of laser fusion cutting.

Laser fusion cutting is applicable to all metals especially stainless steels and other highly alloyed steels, aluminium and titanium alloys. A high quality cut edge is formed but the cutting speeds are relatively low in comparison with active gas cutting mechanisms. The advantage of this process is that the resulting cut edges are free of oxides and have the same corrosion resistance as the substrate. The cut edges may be welded without any post-cutting reparation. The main technical demand is to avoid adherent melt (dross attachment) at the bottom edges of the kerf. A high pressure (above 10 bar) is recommended to remove liquid that can adhere to the underside and solidify as dross.

Laser Oxygen Cutting

The principle of laser oxygen cutting is that the focused laser beam heats the material in an oxidizing atmosphere and ignites an exothermic oxidation reaction of the oxygen with the material. The exothermic reaction supports the laser cutting process by providing additional heat input in the cutting zone resulting into higher cutting speeds compared to laser cutting with inert gases. The laser beam is responsible for igniting and stabilizing a burning process within the kerf, and the assist gas blows out the molten material from the cut zone and protects the laser optics. Figure 16 is a schematic of laser oxygen cutting.

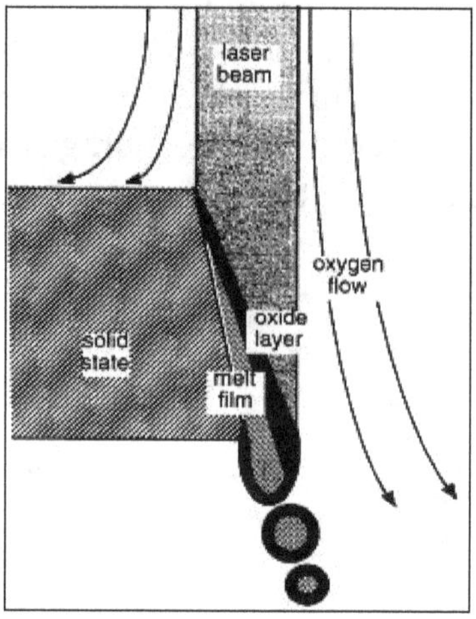

Figure: A sketch of laser oxygen cutting.

Laser oxygen cutting is applicable to mild steel and low-alloyed steel. The formation of the oxide layer on the cutting front increases the absorption of the laser radiation compared to absorption of a pure metallic melt. The oxides reduce the viscosity and surface tension of melt and thereby simplify melt ejection. However, the resulting cut edges are oxidized.

Thermal Stress Cracking

Mechanism of thermal stress cutting is that laser beams heat an area of fragile material to produce evident temperature gradient. The high surface temperature makes expansion and inner lower temperature hinders expansion, forming pulling stress in the surface and radial crushing stress inside. When the two stresses exceed fracture limit strength of the workpiece, crackle appears. And then the work piece is broken along the normal direction of the crack. It is suitable for glasses and ceramics.

Stealth Dicing

In fabrication processes, wafer dicing plays an early and critical role in the quality of the final product. Two conventional methods for dicing wafers — blade dicing and laser ablation dicing – pulverize the wafer material in the cutting path (known as dicing street or kerf). In doing so, these techniques can cause problems such as debris, damage to the device and loss of precious semiconductor material.

A laser technique called stealth dicing can overcome these problems by cutting the wafer from the inside. The method involves two steps: subsurface laser-induced perforations of the wafer followed by tape expansion to separate individual chips.

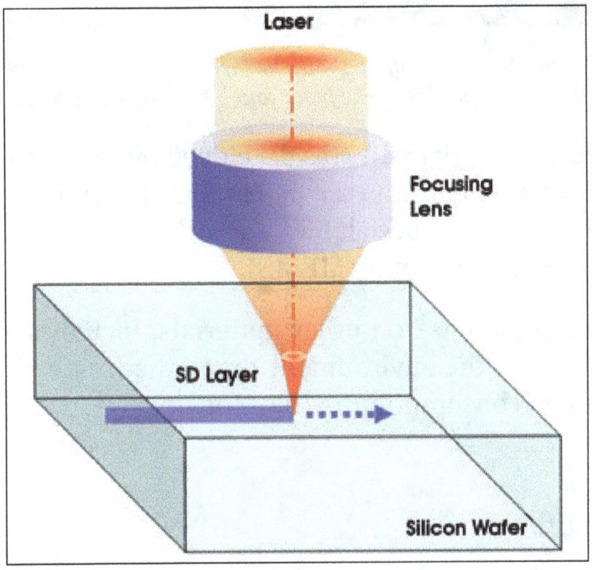

Figure: In stealth dicing, a laser beam that transmits through the wafer is focused beneath the surface, creating a row of perforations in a "stealth dicing" (SD) layer.

The initial subsurface treatment uses a pulsed laser beam at a wavelength that transmits through the wafer. For pure silicon wafers, near-infrared lasers are favored over short-wavelength lasers because of the absorption of shorter wavelengths by silicon. The pulsed laser beam is focused to a point beneath the wafer's surface using an optical system. When the laser beam hits a peak power density at the focal point, the wafer absorbs the energy, and a pinpoint hole is created. The laser also causes compressive and tensile stress near the hole. As the laser beam travels the length of the wafer at a processing speed of 300 mm/s for a 120-µm-thick wafer, it perforates the inner layer of the wafer. The front and back surfaces remain pristine.

After laser processing, the wafer is still in one piece. The separation of individual chips occurs in the next step, when the tape upon which the wafer rests is expanded. The stress areas introduced by laser processing help this separation. A strong compressive stress around the hole is bounded on top and bottom by strong spots of tensile stress, and this sandwich of stress creates fissures along the laser's path – like fault lines splitting the Earth – when the tape expands. Each chip is pulled free from the others. Thus, the cutting of the wafer originates from within the wafer and not from its surface.

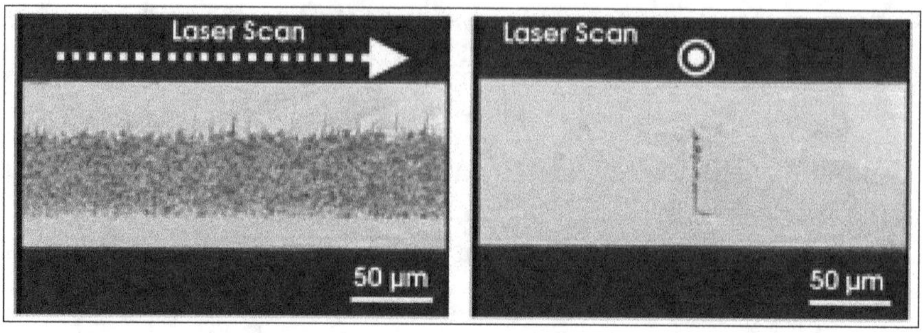

Figure: In the stealth dicing method, the laser beam creates perforations beneath the wafer's surface. The front and back surfaces are not affected.

The stealth dicing technique originally was developed by Hamamatsu Photonics KK in Japan for dicing ultrathin semiconductor wafers, but it has performed well on silicon wafers of all thicknesses and on specialty wafers with die attach films, low-k materials and micro electro mechanical systems (MEMS).

The technique also may find use on other materials, including sapphire and glass. In addition, it is gentler to the environment than other methods because it is a dry process, reducing water consumption by 622 tons per year for each unit using the technology.

Dicing Silicon Wafers

Silicon wafers thicker than 100 µm traditionally are cut by diamond blades or saws, and the blade's thickness, grit size, and rotating and cutting speeds affect the cutting quality. This technique has been refined over the years, yet it still causes problems that cannot be ignored. Chipping of the wafer occurs because of the force of the sawing process. The large cut width (kerf) caused by the blade wastes the wafer and produces debris that must be cleaned up. Other issues include wearing down of the blade over time and the cost of blade replacement.

For silicon wafers less than 100 µm thick, laser ablation offers an alternative to the blade technique, which is too powerful for the delicate thin wafers. However, laser ablation has its own problems. Vaporizing the wafer with a laser along the dicing path creates molten debris and micro cracks. The debris deposited on the surface of the wafer is difficult to clean up, and the cracks result in chips with lower strength.

In contrast, stealth dicing does not generate the problems brought on by either the blade or laser ablation techniques. With stealth dicing, there is no chipping or debris, which eliminates the need for a cleanup process. It also reduces the amount of wasted wafer because of its very narrow dicing path, offering space on the wafer for more chips and higher production yield. Also, the technique does not cause heat damage because the laser cuts below the surface of the wafer; this contributes to rendering the chips more resistant to breakage.

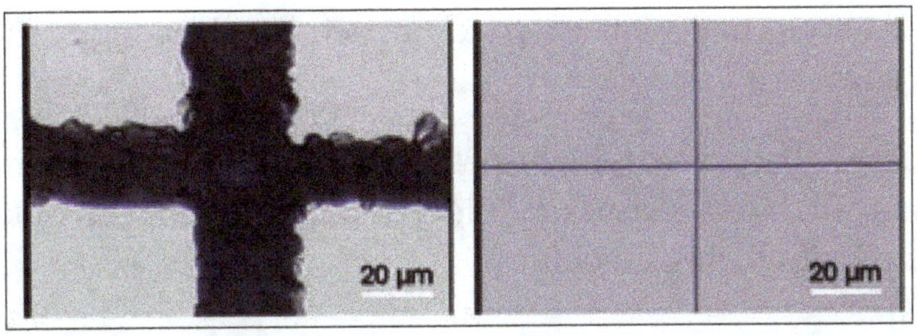

Figure: Chips are diced from a 100-µm-thick silicon wafer by the blade method at a processing speed of 100 mm/s (left) and by the stealth dicing method at a processing speed of 300 mm/s (right).

Dicing Specialty Wafers

Specialty wafers — such as those with die attach films and those made of low-k material — and MEMS devices can pose problems for blade and laser ablation techniques.

Die attach films are used as adhesives for stacking and pasting multiple thin chips in a package. For the conventional blade method, the film adds another level of dicing problems because it is cut at the same time as the wafer; this can cause warping because the wafer and film have different mechanical properties.

However, stealth dicing handles these wafers with no problem. The wafer is laser-processed first, and then the die attach film is attached to the wafer. When the tape expands, the film is separated into the exact shape of the chips because of the very high adhesion forces between the film and wafer.

Low-k materials are widely used in chips to lower the dielectric constant of interlayer insulating films to allow higher-speed operation. For dicing wafers with low-k materials (especially ones that require a completely dry process), conventional dicing processes cause chipping, debris and delamination as well as large kerf widths. Stealth dicing, on the other hand, produces a clean cut.

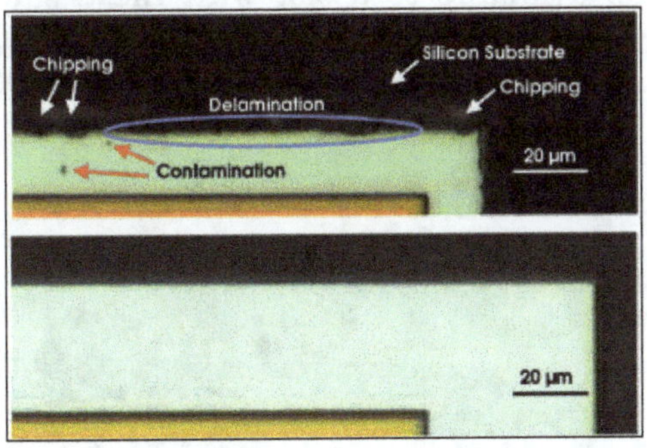

Figure: Low-k device chips are shown being cut by the laser ablation method (top) and by the stealth dicing method (bottom).

MEMS are tiny mechanical electromechanical devices built on a silicon substrate and used in a variety of applications such as biotechnology and communications. These devices are very sensitive to external impacts, which makes stealth dicing an ideal method to use because it has a low external impact. No chipping or cracks occur, and extremely sharp edges are possible.

Stealth dicing offers a dry process, high speed, high quality (no chipping or debris) and low kerf loss. It is available as a laser module that is integrated with optical systems and that is incorporated into dicing machines.

Flame Cutting

The process often called flame cutting is known by many names, such as oxy acetylene Cutting, oxy fuel gas cutting, oxygen burning, steel burning and other terms too numerous to mention. The process is now about 111-112 years old as it was patented in 1901 by Thomas Fletcher. One of the first commercial applications was what the British referred to as "an unauthorized bank entry", or a "safe cracking".

Working

The process is a thermo-chemical process requiring a source of intense heat, referred to as "Preheat", and pure oxygen. The pure oxygen requirement is of a 99.5% minimum purity, which is the minimum purity typically provided in cylinders and bulk systems. A reduction in purity of.1% will reduce cutting speed about 10%, so one can see the necessity for the purity.

While the source of the oxygen may be pure, improper connections, bad hoses, or leaks of any sort can allow impurities in the system thereby reducing the cutting speed. The high purity of the Oxygen presents a very dangerous situation and requires extreme care in the selection of equipment and the design of piping systems for its use. There are numerous documents available to assist with these safety issues. Federal Law (OSHA) presents numerous requirements for the equipment to be used, as well as the procedures to be followed.

The process can be used for cutting/severing many different materials with the basic requirement that the oxide formed must have a lower melting point than the base material to be cut. A good example of a material one cannot cut is Aluminum whose melting point is in the order of 1,200-1,300 degrees F but whose oxide is in the range of 5,000+ degrees F. Stainless steel has additives such as Chromium and Nickel added for "Oxidization Resistance" and this is the process (Oxidization) one is trying to promote, i.e. this type material is not cut without some assistance.

Thick Material can be Cut

The process can cut from sheet metal thicknesses to 100 inch material. Prior to 1958, when Plasma was introduced, all thicknesses were flame cut. Today metal thinner than 3/8-1/2 are typically only cut by plasma or laser, as thin materials are very challenging to flame cut. Thin material requires a very intense preheat as the cutting speed required for a good quality cut is very high, therefore a high temperature high velocity preheat is desired.

The process for all thicknesses is the same and that is the material must be "preheated" to a temperature of 1,600-1,800 degrees F, then the pure Oxygen is discharged into the preheated area and the steel is then oxidized or burned, hence the term "burning". (The term "burner" is often used for the operator.) The quality of the final cut surface can be quite excellent with a sharp top edge, square/flat cut surface, and a sharp slag- free lower edge.

Parameters to Control Flame Cutting

Cut quality is dependent on a number of parameters, the primary one being the competency of the burner/operator. A brief, but not complete list of items/parameters that control cut quality includes:

1. Proper cutting tip selection.
2. Proper cutting oxygen pressure selection.
3. Proper preheat (a function of the cutting tip and the operator) this includes quantity of preheat as well as ratio of Oxygen to the fuel gas.
4. Proper cutting speed.
5. Proper tip to work.
6. Proper selection of the fuel gas.
7. A gas supply system (both fuel gas and Oxygen) that is sufficient to supply the quantity of gas required at the pressure required.

Most of these parameters may vary depending on the material and surface condition of the plate to be cut. All of these items are controlled by the operator and emphasize the importance of a well trained operator as well as a proper and well maintained machine.

References

- Work-theory-laser-cutting-machine-apple-he: linkedin.com, Retrieved 30,June 2020
- How-laser-cutting-works, application-cutting: spilasers.com, Retrieved 13, July 2020
- What-is-laser-cutting: wisegeek.com, Retrieved 27, May 2020
- What-is-flame-cutting: esabna.com, Retrieved 25, June 2020
- Laser-Dicing-Technique-Cuts-Wafers-from-the-31907: photonics.com, Retrieved 15,March 2020

5

Applications of Lasers in Diverse Fields

Lasers have a wide variety of uses in various industries and fields. A few of these are optical disk drive, laser printing, 3D scanner, laser surgery, fiber-optic communication and laser pointer. This chapter closely examines the applications of lasers in these fields to provide an extensive understanding of the subject.

Optical Disc Drive

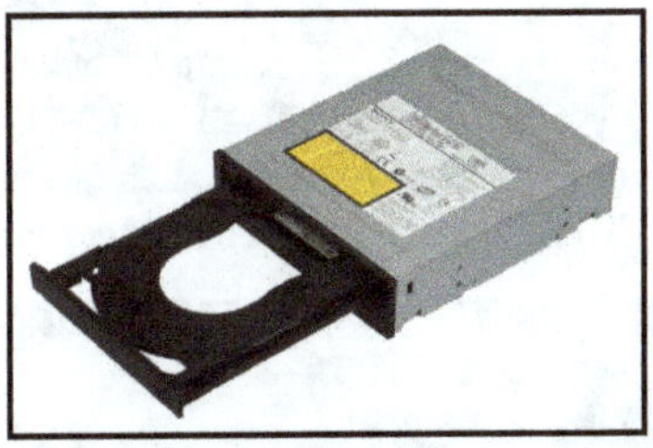

A CD/DVD-ROM computer drive.

In computing, an optical disc drive (ODD) is a disk drive that uses laser light or electromagnetic waves within or near the visible light spectrum as part of the process of reading or writing data to or from optical discs. Some drives can only read from certain discs, but recent drives can both read and record, also called burners or writers. Compact discs, DVDs, and Blu-ray discs are common types of optical media which can be read and recorded by such drives. Optical disc drives that are no longer in production include CD-ROM drive, CD writer drive, and combo (CD-RW/DVD-ROM) drive. As of 2015, DVD writer drive is the most common for desktop PCs and laptops. There are also the DVD-ROM drive, BD-ROM drive, Blu-ray Disc combo (BD-ROM/DVD±RW/CD-RW) drive, and Blu-ray Disc writer drive.

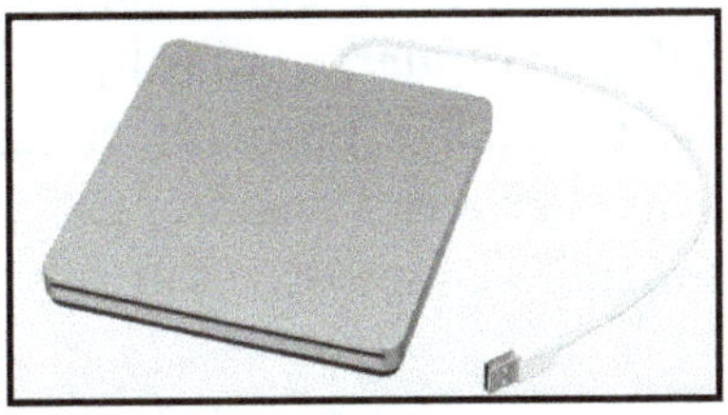

An external Apple USB SuperDrive.

Optical disc drives are an integral part of standalone appliances such as CD players, VCD players, DVD players, Blu-ray disc players, DVD recorders, certain desktop video game consoles, such as Sony PlayStation 4, Microsoft Xbox One, Nintendo Wii U, and Sony PlayStation 3, and certain portable video game consoles, such as Sony PlayStation Portable. They are also very commonly used in computers to read software and consumer media distributed on disc, and to record discs for archival and data exchange purposes. Floppy disk drives, with capacity of 1.44 MB, have been made obsolete: optical media are cheap and have vastly higher capacity to handle the large files used since the days of floppy discs, and the vast majority of computers and much consumer entertainment hardware have optical writers. USB flash drives, high-capacity, small, and inexpensive, are suitable where read/write capability is required.

The CD/DVD drive lens on an Acer laptop.

A Blu-ray drive holds independent lenses for Blu-ray and DVD/CD media. Pictured are lenses from a Blu-ray writer in a Sony Vaio E series laptop.

Disc recording is restricted to storing files playable on consumer appliances (films, music, etc.), relatively small volumes of data (e.g., a standard DVD holds 4.7 gigabytes) for local use, and data for distribution, but only on a small scale; mass-producing large numbers of identical discs is cheaper and faster than individual recording.

Optical discs are used to back up relatively small volumes of data, but backing up of entire hard drives, which as of 2015[update] typically contain many hundreds of gigabytes or even multiple terabytes, is less practical. Large backups are often instead made on external hard drives, as their price has dropped to a level making this viable; in professional environments magnetic tape drives are also used.

History

The first laser disc, demonstrated in 1972, was the *Laservision* 12-inch video disc. The video signal was stored as an analog format like a video cassette. The first digitally recorded optical disc was a 5-inch audio compact disc (CD) in a read-only format created by Philips and Sony in 1975. Five years later, the same two companies introduced a digital storage solution for computers using this same CD size called a CD-ROM. Not until 1987 did Sony demonstrate the erasable and rewritable 5.25-inch optical drive.

Key Components

Laser And Optics

The most important part of an optical disc drive is an *optical path, placed in a pick-up head (PUH),* usually consisting of semiconductor laser, a lens for guiding the laser beam, and photodiodes detecting the light reflection from disc's surface.

Initially, CD lasers with a wavelength of 780 nm were used, being within infrared range. For DVDs, the wavelength was reduced to 650 nm (red color), and the wavelength for Blu-ray Disc was reduced to 405 nm (violet color).

Two main servomechanisms are used, the first one to maintain a correct distance between lens and disc, and ensure the laser beam is focused on a small *laser spot on the disc. The second servo moves a head along the disc's radius, keeping the beam on a groove, a continuous spiral data path.*

The optical sensor out of a CD/DVD drive. The two bigger rectangles are the photodiodes for pits, the inner one for land. This one also includes amplification and minor processing.

On *read only media (ROM), during the manufacturing process the groove, made of pits, is pressed on a flat surface, called land. Because the depth of the pits is approximately one-quarter to one-sixth of the laser's wavelength, the reflected beam's phase is shifted in relation to the incoming reading beam, causing mutual destructive* interference and reducing the reflected beam's intensity. This is detected by photodiodes that output electrical signals.

A recorder encodes (or *burns) data onto a recordable* CD-R, DVD-R, DVD+R, or BD-R disc (called a *blank) by selectively heating parts of an organic* dye layer with a laser. This changes the reflectivity of the dye, thereby creating marks that can be read like the pits and lands on pressed discs. For recordable discs, the process is permanent and the media can be written to only once. While the reading laser is usually not stronger than 5 mW, the writing laser is considerably more powerful. The higher the writing speed, the less time a laser has to heat a point on the media, thus its power has to increase proportionally. DVD burners' lasers often peak at about 200 mW, either in continuous wave and pulses, although some have been driven up to 400 mW before the diode fails.

For rewritable CD-RW, DVD-RW, DVD+RW, DVD-RAM, or BD-RE media, the laser is used to melt a crystalline metal alloy in the recording layer of the disc. Depending on the amount of power applied, the substance may be allowed to melt back (change the phase back) into crystalline form or left in an amorphous form, enabling marks of varying reflectivity to be created.

Double-sided media may be used, but they are not easily accessed with a standard drive, as they must be physically turned over to access the data on the other side.

Double layer (DL) media have two independent data layers separated by a semi-reflective layer. Both layers are accessible from the same side, but require the optics to change the laser's focus. Traditional single layer (SL) writable media are produced with a spiral groove molded in the protective polycarbonate layer (not in the data recording layer), to lead and synchronize the speed of recording head. Double-layered writable media have: a first polycarbonate layer with a (shallow) groove, a first data layer, a semi-reflective layer, a second (spacer) polycarbonate layer with another (deep) groove, and a second data layer. The first groove spiral usually starts on the inner edge and extends outwards, while the second groove starts on the outer edge and extends inwards.

Some drives support Hewlett-Packard's LightScribe photothermal printing technology for labeling specially coated discs.

Rotational Mechanism

The rotational mechanism in an optical drive differs considerably from that of a hard disk drive's, in that the latter keeps a constant angular velocity (CAV), in other words a constant number of revolutions per minute (RPM). With CAV, a higher throughput is generally achievable at the outer disc compared to the inner.

On the other hand, optical drives were developed with an assumption of achieving a constant throughput, in CD drives initially equal to 150 KiB/s. It was a feature important for streaming audio data that always tend to require a constant bit rate. But to ensure no disc capacity was wasted, a head had to transfer data at a maximum linear rate at all times too, without slowing on the outer rim of disc. This led to optical drives—until recently—operating with a constant linear velocity (CLV). The spiral *groove of the disc passed under its head at a constant speed. The implication of CLV, as opposed to CAV, is that disc angular velocity is no longer constant, and the spindle motor needed to be designed to vary its speed from between 200 RPM on the outer rim and 500 RPM on the inner.*

Later CD drives kept the CLV paradigm, but evolved to achieve higher rotational speeds, popularly described in multiples of a base speed. As a result, a 4× drive, for instance, would rotate at 800-2000 RPM, while transferring data steadily at 600 KiB/s, which is equal to 4 × 150 KiB/s.

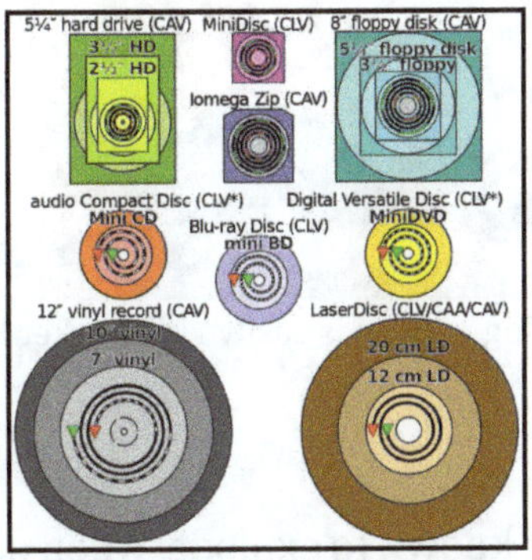

Comparison of several forms of disk storage showing tracks (not-to-scale); green denotes start and red denotes end.* Some CD-R(W) and DVD-R(W)/DVD+R(W) recorders operate in ZCLV, CAA or CAV. modes.

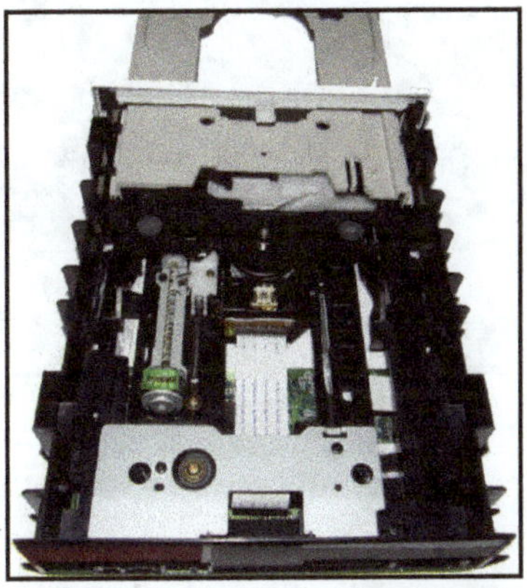

A CD-ROM Drive (without case).

For DVDs, base or 1× speed is 1.385 MB/s, equal to 1.32 MiB/s, approximately 9 times faster than the CD base speed. For Blu-ray drives, base speed is 6.74 MB/s, equal to 6.43 MiB/s.

Because keeping a constant transfer rate for the whole disc is not so important in most contemporary CD uses, a pure CLV approach had to be abandoned to keep the rotational speed of the disc safely low while maximizing data rate. Some drives work in a partial CLV (PCLV) scheme, by switching from CLV to CAV only when a rotational limit is reached. But switching to CAV requires considerable changes in hardware design, so instead most drives use the zoned constant linear velocity (Z-CLV) scheme. This divides the disc into several zones, each having its own constant linear velocity. A Z-CLV recorder rated at "52×", for example, would write at 20× on the innermost zone and then progressively increase the speed in several discrete steps up to 52× at the outer rim. Without higher rotational speeds, increased read performance may be attainable by simultaneously reading more than one point of a data groove, but drives with such mechanisms are more expensive, less compatible, and very uncommon.

The Z-CLV recording pattern is easily visible after burning a DVD-R.

Limit

Both DVDs and CDs have been known to explode when damaged and/or spun at excessive speed. This imposes a constraint on the maximum speed (56× for CDs or around 18× in the case of DVDs) at which drives can operate.

An exploded disc.

Loading Mechanisms

Current optical drives use either a *tray-loading mechanism, where the disc is loaded onto a motorized or manually operated tray, or a slot-loading mechanism, where the disc is slid into a slot and drawn in by motorized rollers. With both types of mechanism, if a CD or DVD is left in the drive after the computer is turned off, the disc cannot be ejected using the normal eject mechanism of the drive. However, tray-loading drives account for this situation by providing a small hole where one can insert a straightened paperclip to manually open the drive tray to retrieve the disc. Slot-loading optical disc drives have the disadvantages that they cannot usually accept the smaller* 80 mm discs (unless 80 mm optical disc adapter is used) or any non-standard sizes, usually have no emergency eject hole or eject button, and therefore have to be disassembled if the optical disc cannot be ejected normally. However, the Nintendo Wii, because of backwards compatibility with Nintendo GameCube games, and PlayStation 3 video game consoles are able to load standard size DVDs and 80 mm discs in the same slot-loading drive.

A small number of drive models, mostly compact portable units, have a *top-loading mechanism where the drive lid is opened upwards and the disc is placed directly onto the spindle* (for example, all PlayStation One consoles, portable CD players, and some standalone CD recorders all feature top-loading drives). These sometimes have the advantage of using spring-loaded ball bearings to hold the disc in place, minimizing damage to the disc if the drive is moved while it is spun up.

Some early CD-ROM drives used a mechanism where CDs had to be inserted into special cartridges or caddies, somewhat similar in appearance to a 3.5" floppy diskette. This was intended to protect the disc from accidental damage by enclosing it in a tougher plastic casing, but did not gain wide acceptance due to the additional cost and compatibility concerns—such drives would also inconveniently require "bare" discs to be manually inserted into an openable caddy before use. Ultra Density Optical and Universal Media Disc use optical disc cartridges.

There were also some early CD-ROM drives for desktop PCs in which its tray-loading mechanism will eject slightly and user has to pull out the tray manually to load CD, similar to the tray ejecting method used in internal optical disc drives of modern laptops and modern external slim portable optical disc drives. Like the top-loading mechanism, they have spring-loaded ball bearings on the spindle.

Computer Interfaces

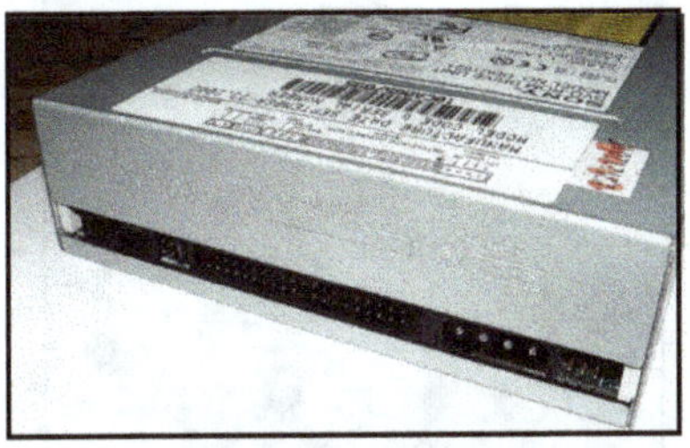

Digital audio output, analog audio output, and parallel ATA interface.

Most internal drives for personal computers, servers and workstations are designed to fit in a standard 5.25" drive bay and connect to their host via an ATA or SATA interface. Additionally, there may be digital and analog outputs for audio. The outputs may be connected via a header cable to the sound card or the motherboard. At one time, computer software resembling cd players controlled playback of the CD. Today the information is extracted from the disc as data, to be played back or converted to other file formats.

External drives usually have USB or FireWire interfaces. Some portable versions for laptops power themselves from batteries or directly from their interface bus.

Drives with SCSI interface were made, but they are less common and tend to be more expensive, because of the cost of their interface chipsets, more complex SCSI connectors, and small volume of sales.

When the optical disc drive was first developed, it was not easy to add to computer systems. Some computers such as the IBM PS/2 were standardizing on the 3.5" floppy and 3.5" hard disk, and did not include a place for a large internal device. Also IBM PCs and clones at first only included a single (parallel) ATA drive interface, which by the time the CDROM was introduced, was already being used to support two hard drives. Early laptops simply had no built-in high-speed interface for supporting an external storage device.

This was solved through several techniques:

- Early sound cards could include a CD-ROM drive interface. Initially, such interfaces were proprietary to each CD-ROM manufacturer. A sound card could often have two or three different interfaces which are able to communicate with cdrom drive.
- A parallel port external drive was developed that connected between a printer and the computer. This was slow but an option for laptops.
- A PCMCIA optical drive interface was also developed for laptops.
- A SCSI card could be installed in desktop PCs for an external SCSI drive enclosure, though SCSI was typically much more expensive than other options.

Internal Mechanism of a Drive

Internal mechanism of a DVD-ROM Drive.

The optical drives in the photos are shown right side up; the disc would sit on top of them. The laser and optical system scans the underside of the disc.

With reference to the top photo, just to the right of image center is the disc motor, a metal cylinder, with a gray centering hub and black rubber drive ring on top. There is a disc-shaped round clamp, loosely held inside the cover and free to rotate; it's not in the photo. After the disc tray stops moving inward, as the motor and its attached parts rise, a magnet near the top of the rotating assembly contacts and strongly attracts the clamp to hold and center the disc. This motor is an "outrunner"-style brushless DC motor which has an external rotor – every visible part of it spins.

Two parallel guide rods that run between upper left and lower right in the photo carry the "sled", the moving optical read-write head. As shown, this "sled" is close to, or at the position where it reads or writes at the edge of the disc. To move the "sled" during continuous read or write operations, a stepper motor rotates a leadscrew to move the "sled" throughout its total travel range. The motor, itself, is the short gray cylinder just to the left of the most-distant shock mount; its shaft is parallel to the support rods. The leadscrew is the rod with evenly-spaced darker details; these are the helical grooves that engage a pin on the "sled".

In contrast, the mechanism shown in the second photo, which comes from a cheaply made DVD player, uses less accurate and less efficient brushed DC motors to both move the sled and spin the disc. Some older drives use a DC motor to move the sled, but also have a magnetic rotary encoder to keep track of the position. Most drives in computers use stepper motors.

The gray metal chassis is shock-mounted at its four corners to reduce sensitivity to external shocks, and to reduce drive noise from residual imbalance when running fast. The soft shock mount grommets are just below the brass-colored screws at the four corners (the left one is obscured).

In the third photo, the components under the cover of the lens mechanism are visible. The two permanent magnets on either side of the lens holder as well as the coils that move the lens can be seen. This allows the lens to be moved up, down, forwards, and backwards to stabilize the focus of the beam.

In the fourth photo, the inside of the optics package can be seen. Note that since this is a CD-ROM drive, there is only one laser, which is the black component mounted to the bottom left of the assembly. Just above the laser are the first focusing lens and prism that direct the beam at the disc. The tall, thin object in the center is a half-silvered mirror that splits the laser beam in multiple directions. To the bottom right of the mirror is the main photodiode that senses the beam reflected off the disc. Above the main photodiode is a second photodiode that is used to sense and regulate the power of the laser.

The irregular orange material is flexible etched copper foil supported by thin sheet plastic; these are "flexible printed circuits" that connect everything to the electronics (which is not shown).

Compatibility

Most optical drives are backwards compatible with their ancestors up to CD, although this is not required by standards.

Compared to a CD's 1.2 mm layer of polycarbonate, a DVD's laser beam only has to penetrate 0.6 mm in order to reach the recording surface. This allows a DVD drive to focus the beam on a smaller spot size and to read smaller pits. DVD lens supports a different focus for CD or DVD media with same laser. With the newer Blu-ray disc drives, the laser only has to penetrate 0.1 mm of material. Thus the optical assembly would normally have to have an even greater focus range. In practice, the Blu-ray optical system is separate from the DVD/CD system.

Optical disc drive	Optical disc or optical media													
	Pressed CD	CD-R	CD-RW	Pressed DVD	DVD-R	DVD +R	DVD-RW	DVD +RW	DVD+R DL	Pressed CAT BD	BD-R	BD-RE	BD-R DL	BD-RE DL
Audio CD player	Read	Read [1]	Read [2]	None	None	None	None	None	None	None	None	None	None	None
CD-ROM drive	Read	Read [1]	Read [2]	None	None	None	None	None	None	None	None	None	None	None
CD-R re-corder	Read	Write	Read	None	None	None	None	None	None	None	None	None	None	None
CD-RW recorder	Read	Write	Write	None	None	None	None	None	None	None	None	None	None	None
DVD-ROM drive	Read	Read [3]	Read [3]	Read	Read [4]	Read [4]	Read [4]	Read [4]	Read [5]	None	None	None	None	None
DVD-R recorder	Read	Write	Write	Read	Write	Read [6]	Read	Read [6]	Read [5]	None	None	None	None	None
DVD-RW recorder	Read	Write	Write	Read	Write	Read [7]	Write [8]	Read [6]	Read [5]	None	None	None	None	None
DVD+RW recorder	Read	Write	Write	Read	Read [6]	Read [9]	Read [6]	Write	Read [5]	None	None	None	None	None
DVD+R recorder	Read	Write	Write	Read	Read [6]	Write	Read [6]	Write	Read [5]	None	None	None	None	None
DVD±RW recorder	Read	Write	Write	Read	Write	Write	Write	Write	Read [5]	None	None	None	None	None
DVD±RW/ DVD+R DL recorder	Read	Write	Write	Read	Write [10]	Write	Write [10]	Write	Write	None	None	None	None	None
BD-ROM	Read	Read	Read	Read	Read	Read	Read	Read	Read	Read	Read	Read	Read	Read
BD-R re-corder	Read [11]	Write [11]	Write [11]	Read	Write	Write	Write	Write	Write	Read	Write	Read	Read	Read
BD-RE recorder	Read [11]	Write [11]	Write [11]	Read	Write	Write	Write	Write	Write	Read	Write	Write	Read	Read
BD-R DL recorder	Read [11]	Write [11]	Write [11]	Read	Write	Write	Write	Write	Write	Read	Write	Read	Write	Read
BD-RE DL recorder	Read [11]	Write [11]	Write [11]	Read	Write	Write	Write	Write	Write	Read	Write	Write	Write	Write

- Some types of CD-R media with less-reflective dyes may cause problems.
- May not work in non MultiRead-compliant drives.
- May not work in some early-model DVD-ROM drives. CD-R would not work in any drive that did not have a 780 nm laser. CD-RW compatibility varied.
- DVD+RW discs did not work in early video players that played DVD-RW discs. This was not due to any incompatibility with the format but was a deliberate feature built into the firmware by one drive manufacturer.
- Read compatibility with existing DVD drives may vary greatly with the brand of DVD+R DL media used. Also drives that predated the media did not have the book code for DVD+R DL media in their firmware (this was not an issue for DVD-R DL though some drives could only read the first layer).
- Early DVD+RW and DVD+R recorders could not write to DVD-R(W) media (and vice versa).
- Will work in all drives that read DVD-R as compatibility ID byte is the same.
- Recorder firmware may blacklist or otherwise refuse to record to some brands of DVD-RW media.
- DVD+RW format was released before DVD+R. All DVD+RW only drives could be upgraded to write DVD+R discs by a firmware upgrade.
- As of April 2005, all DVD+R DL recorders on the market are Super Multi-capable.
- As of October 2006, recently released BD drives are able to read and write CD media.

Recording Performance

During the times of CD writer drives, they are often marked with three different speed ratings. In these cases, the first speed is for write-once (R) operations, the second speed for re-write (RW) operations, and the last speed for read-only (ROM) operations. For example, a 40×/16×/48× CD writer drive is capable of writing to CD-R media at speed of 40× (6,000 KB/s), writing to CD-RW media at speed of 16× (2,400 KB/s), and reading from a CD-ROM media at speed of 48× (7,200 KB/s).

During the times of combo (CD-RW/DVD-ROM) drives, an additional speed rating (e.g., the 16× in 52×/32×/52×/16×) is designated for DVD-ROM media reading operations.

For DVD writer drives, Blu-ray disc combo drives, and Blu-ray disc writer drives, the writing and reading speed of their respective optical media are specified in its retail box, user's manual, and/or bundled brochures or pamphlets.

In the late 1990s, *buffer underruns became a very common problem as high-speed CD recorders began to appear in home and office computers, which—for a variety of reasons—often could not muster the I/O performance to keep the data stream to the recorder steadily fed. The recorder, should it run short, would be forced to halt the recording process, leaving a truncated track that usually renders the disc useless.*

In response, manufacturers of CD recorders began shipping drives with "buffer underrun protection" (under various trade names, such as Sanyo's "BURN-Proof", Ricoh's "JustLink" and Yamaha's "Lossless Link"). These can suspend and resume the recording process in such a way that the gap the stoppage produces can be dealt with by the error-correcting logic built into CD players and CD-ROM drives. The first of these drives were rated at 12× and 16×.

While drives are burning DVD+R, DVD+RW and all Blu-ray formats, they do not require any such error correcting recovery as the recorder is able to place the new data exactly on the end of the suspended write effectively producing a continuous track (this is what the DVD+ technology achieved). Although later interfaces were able to stream data at the required speed, many drives now write in a 'zoned constant linear velocity'. This means that the drive has to temporarily suspend the write operation while it changes speed and then recommence it once the new speed is attained. This is handled in the same manner as a buffer underrun.

The internal buffer of optical disc writer drives is: 8 MiB when recording BD-R/BD-R DL/BD-RE/BD-RE DL media; 2 MiB when recording DVD-R/DVD-RW/DVD-R DL/ DVD+R/DVD+RW/DVD+RW DL/DVD-RAM/CD-R/CD-RW media.

Recording Schemes

CD recording on personal computers was originally a batch-oriented task in that it required specialised authoring software to create an "image" of the data to record, and to record it to disc in the one session. This was acceptable for archival purposes, but limited the general convenience of CD-R and CD-RW discs as a removable storage medium.

Packet writing is a scheme in which the recorder writes incrementally to disc in short bursts, or packets. Sequential packet writing fills the disc with packets from bottom up. To make it readable in CD-ROM and DVD-ROM drives, the disc can be *closed at any time by writing a final table-of-contents to the start of the disc; thereafter, the disc cannot be packet-written any further. Packet writing, together with support from the* operating system and a file system like UDF, can be used to mimic random write-access as in media like flash memory and magnetic disks.

Fixed-length packet writing (on CD-RW and DVD-RW media) divides up the disc into padded, fixed-size packets. The padding reduces the capacity of the disc, but allows the recorder to start and stop recording on an individual packet without affecting its neighbours. These resemble the block-writable access offered by magnetic media closely enough that many conventional file systems will work as-is. Such discs, however, are not readable in most CD-ROM and DVD-ROM drives or on most operating systems without additional third-party drivers. The division into packets is not as reliable as it may seem as CD-R(W) and DVD-R(W) drives can only locate data to within a data block. Although generous gaps (the padding referred to above) are left between blocks, the drive nevertheless can occasionally miss and either destroy some existing data or even render the disc unreadable.

The DVD+RW disc format eliminates this unreliability by embedding more accurate timing hints in the data groove of the disc and allowing individual data blocks (or even bytes) to be replaced without affecting backwards compatibility (a feature dubbed "lossless linking"). The format itself was designed to deal with discontinuous recording because it was expected to be widely used in digital video recorders. Many such DVRs use variable-rate video compression schemes which require them to record in short bursts; some allow simultaneous playback and recording by alternating quickly between recording to the tail of the disc whilst reading from elsewhere. The Blu-ray disc system also encompasses this technology.

Mount Rainier aims to make packet-written CD-RW and DVD+RW discs as convenient to use as that of removable magnetic media by having the firmware format new discs in the background and manage media defects (by automatically mapping parts of the disc which have been worn out by erase cycles to reserve space elsewhere on the disc). As of February 2007, support for Mount Rainier is natively supported in Windows Vista. All previous versions of Windows require a third-party solution, as does Mac OS X.

Recorder Unique Identifier

Owing to pressure from the music industry, as represented by the IFPI and RIAA, Philips developed the *Recorder Identification Code (RID) to allow media to be uniquely associated with the recorder that has written it. This standard is contained in the* Rainbow Books. The RID-Code consists of a supplier code (e.g. "PHI" for Philips), a model number and the unique ID of the recorder. Quoting Philips, the RID "enables a trace for each disc back to the exact machine on which it was made using coded information in the recording itself. The use of the RID code is mandatory."

Although the RID was introduced for music and video industry purposes, the RID is included on every disc written by every drive, including data and backup discs. The value of the RID is questionable as it is (currently) impossible to locate any individual recorder due to there being no database.

Source IDentification Code

The Source IDentification Code (SID) is an eight character supplier code that is placed on optical discs by the manufacturer. The SID identifies not only manufacturer, but also the individual factory and machine that produced the disc.

According to Phillips, the administrator of the SID codes, the SID code provides an optical disc production facility with the means to identify all discs mastered and/or replicated in its plant, including the specific Laser Beam Recorder (LBR) signal processor or mould that produced a particular stamper or disc.

Use of RID and SID together in forensics

The standard use of RID and SID mean that each disc written contains a record of the machine that produced a disc (the SID), and which drive wrote it (the RID). This combined knowledge may be very useful to law enforcement, to investigative agencies, and to private and/or corporate investigators.

Laser Printing

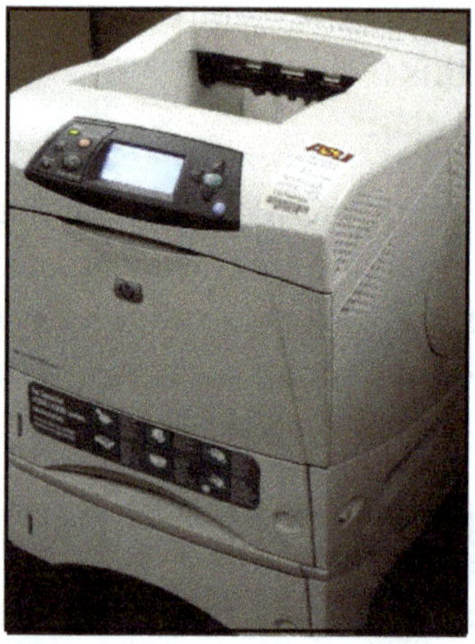

HP LaserJet 4200 series printer, installed atop high-capacity paper feeder.

Laser printing is an electrostatic digital printing process. It produces high-quality text and graphics (and moderate-quality photographs) by repeatedly passing a laser beam back and forth over a negatively charged cylinder called a "drum" to define a differentially charged image. The drum then selectively collects electrically charged powdered ink (toner), and transfers the image to paper, which is then heated in order to permanently fuse the text and/or imagery. As with digital photocopiers and multifunction/all-in-one inkjet printers, laser printers employ a xerographic printing process. However, laser printing differs from analog photocopiers in that the image is produced by the direct scanning of the medium across the printer's photoreceptor. This enables laser printing to copy images more quickly than most photocopiers.

Invented at Xerox PARC in the 1970s, laser printers were introduced for the office and then home markets in subsequent years by IBM, Canon, Xerox, Apple, Hewlett-Packard and many others. Over the decades, quality and speed have increased as price has fallen, and the once cutting-edge printing devices are now ubiquitous.

History

Gary Starkweather invented the laser printer (2009 photo).

In the 1960s, the Xerox Corporation held a dominant position in the photocopier market. In 1969, Gary Starkweather, who worked in Xerox's product development department, had the idea of using a laser beam to "draw" an image of what was to be copied directly onto the copier drum. After transferring to the recently formed Palo Alto Research Center (Xerox PARC) in 1971, Starkweather adapted a Xerox 7000 copier to create SLOT (Scanned Laser Output Terminal). In 1972, Starkweather worked with Butler Lampson and Ronald Rider to add a control system and character generator, resulting in a printer called EARS (Ethernet, Alto Research character generator, Scanned laser output terminal) -- which later became the Xerox 9700 laser printer.

The first commercial implementation of a laser printer was the IBM 3800 in 1976. It was designed for data centers, where it replaced line printers attached to mainframe computers. The IBM 3800 was used for high-volume printing on continuous stationery, and achieved speeds of 215 pages per minute (ppm), at a resolution of 240 dots per inch (dpi). Over 8,000 of these printers were sold. The Xerox 9700 was brought to market in 1977. Unlike the IBM 3800, the Xerox 9700 was not targeted to replace any particular existing printers; but, it did have limited support for the loading of fonts. The Xerox 9700 excelled at printing high-value documents on cut-sheet paper with varying content (e.g. insurance policies).

In 1979, inspired by the Xerox 9700's commercial success, Japanese camera and optics company, Canon, developed a low-cost, *desktop laser printer: the Canon LBP-10. Canon then began work on a much-improved print engine, the Canon CX, resulting in the LBP-CX printer. Lacking experience in selling to computer users, Canon sought partnerships with three* Silicon Valley companies: Diablo Data Systems (who turned them down), Hewlett-Packard (HP), and Apple Computer.

The first laser printer designed for office use reached market in 1981: the Xerox Star 8010. The system used a desktop metaphor that was unsurpassed in commercial sales, until the Apple Macintosh. Although it was innovative, the Star workstation was a prohibitively expensive (US$17,000) system, affordable only to a fraction of the businesses and institutions at which it was targeted.

The first laser printer intended for mass-market sales was the HP LaserJet, released in 1984; it used the Canon CX engine, controlled by HP software. The LaserJet was quickly followed by printers from Brother Industries, IBM, and others. First-generation machines had large photosensitive drums, of circumference greater than the loaded paper's length. Once faster-recovery coatings were developed, the drums could touch the paper multiple times in a pass, and therefore be smaller in diameter.

In 1985, Apple introduced the LaserWriter (also based on the Canon CX engine), but used the newly released PostScript page-description language. Up until this point, each manufacturer used its own proprietary page-description language, making the supporting software complex and expensive. PostScript allowed the use of text, fonts, graphics, images, and color largely independent of the printer's brand or resolution. PageMaker, written by Aldus for the Macintosh and LaserWriter, was also released in 1985 and the combination became very popular for desktop publishing.:13/23:364 Laser printers brought exceptionally fast and high-quality text printing in multiple fonts on a page, to the business and consumer markets. No other commonly available printer during this era could also offer this combination of features.

Printing Process

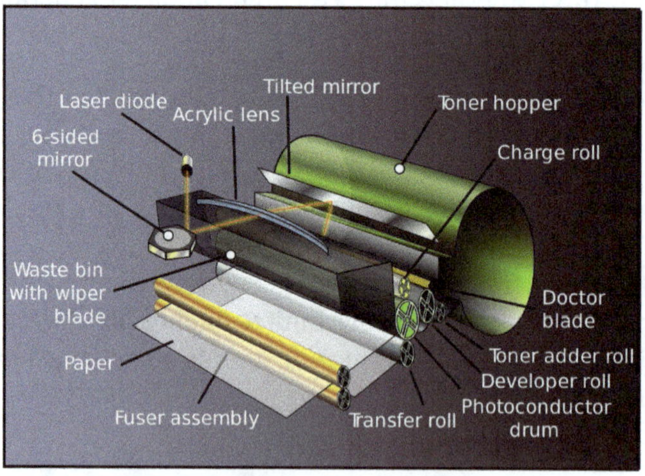

Diagram of a laser printer.

A laser beam (typically, an aluminium gallium arsenide (AlGaAs) semiconductor laser) projects an image of the page to be printed onto an electrically charged, selenium-coated, rotating, cylindrical drum (or, more commonly in subsequent versions, a drum called an organic photoconductor made of N-vinylcarbazole, an organic monomer). Photoconductivity allows the charged electrons to fall away from the areas exposed to light. Powdered ink (toner) particles are then electrostatically attracted to the charged areas of the drum that have not been laser-beamed. The drum then transfers the image onto paper (which is passed through the machine) by direct contact. Finally the paper is passed onto a finisher, which uses intense heat to instantly fuse the toner/image onto the paper.

There are typically seven steps involved in the process:

Raster Image Processing

The document to be printed is encoded in a page description language such as PostScript, Printer Command Language (PCL), or Open XML Paper Specification (OpenXPS). The raster image processor converts the page description into a bitmap which is stored in the printer's raster memory. Each horizontal strip of dots across the page is known as a raster line or scan line.

Laser printing differs from other printing technologies in that each page is always rendered in a single continuous process without any pausing in the middle, while other technologies like inkjet can pause every few lines. To avoid a buffer underrun (where the laser reaches a point on the page before it has the dots to draw there), a laser printer typically needs enough raster memory to hold the bitmap image of an entire page.

Memory requirements increase with the square of the dots per inch, so 600 dpi requires a minimum of 4 megabytes for monochrome, and 16 megabytes for color at 600 dpi. For fully graphical output using a page description language, a minimum of 1 megabyte of memory is needed to store an entire monochrome letter/A4 sized page of dots at 300 dpi. At 300 dpi, there are 90,000 dots per square inch (300 dots per linear inch). A typical 8.5 × 11 sheet of paper has 0.25-inch (6.4 mm) margins, reducing the printable area to 8.0 by 10.5 inches (200 mm × 270 mm), or 84 square inches. 84 sq/in × 90,000 dots per sq/in = 7,560,000 dots. 1 megabyte = 1,048,576 bytes, or 8,388,608 bits, which is just large enough to hold the entire page at 300 dpi, leaving about 100 kilobytes to spare for use by the raster image processor.

In a color printer, each of the four CMYK toner layers is stored as a separate bitmap, and all four layers are typically preprocessed before printing begins, so a minimum of 4 megabytes is needed for a full-color letter-size page at 300 dpi.

During the 1980s, memory chips were still very expensive, which is why entry-level laser printers in that era always came with four-digit suggested retail prices in US dollars. Memory prices later plunged, and 1200 dpi printers have been widely available in the consumer market since 2008. 2400 dpi electrophotographic printing plate makers, essentially laser printers that print on plastic sheets, are also available.

Charging

In older printers, a corona wire positioned parallel to the drum or, in more recent printers, a primary charge roller, projects an electrostatic charge onto the photoreceptor (otherwise named the photo conductor unit), a revolving photosensitive drum or belt, which is capable of holding an electrostatic charge on its surface while it is in the dark.

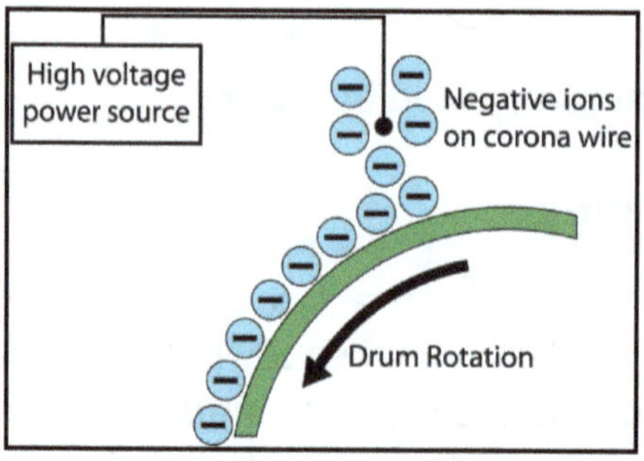

Applying a negative charge to the photosensitive drum.

An AC bias voltage is applied to the primary charge roller to remove any residual charges left by previous images. The roller will also apply a DC bias on the drum surface to ensure a uniform negative potential.

Numerous patents[*specify*] describe the photosensitive drum coating as a silicon sandwich with a photocharging layer, a charge leakage barrier layer, as well as a surface layer. One version[*specify*] uses amorphous silicon containing hydrogen as the light receiving layer, Boron nitride as a charge leakage barrier layer, as well as a surface layer of doped silicon, notably silicon with oxygen or nitrogen which at sufficient concentration resembles machining silicon nitride.

Exposing

A laser printer uses a laser because lasers are able to form highly focused, precise, and intense beams of light, especially over the short distances inside of a printer. The laser is aimed at a rotating polygonal mirror which directs the light beam through a system of lenses and mirrors onto the photoreceptor drum, writing pixels at rates up to sixty five million times per second. The drum continues to rotate during the sweep, and the angle of sweep is canted very slightly to compensate for this motion. The stream of rasterized data held in the printer's memory rapidly turns the laser on and off as it sweeps.

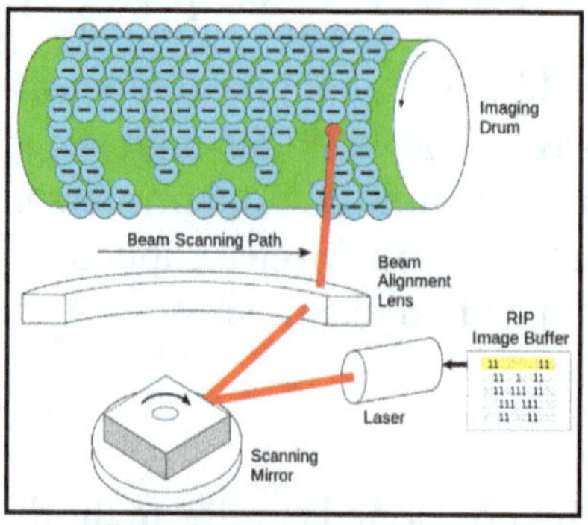

Laser light selectively neutralizes the negative charge on the photoreceptive drum, to form an electrostatic image.

Laser unit from a Dell P1500. The white hexagon is the rotating scanner mirror.

The laser beam neutralizes (or reverses) the charge on the surface of the drum, leaving a static electric negative image on the drum's surface which will repel the negatively charged toner particles. The areas on the drum which were struck by the laser, however, momentarily have no charge, and the toner being pressed against the drum by the toner-coated developer roll in the next step moves from the roll's rubber surface to the uncharged portions of the surface of the drum.

Some non-laser printers (LED printers) use an array of light emitting diodes spanning the width of the page to generate an image, rather than using a laser. "Exposing" is also known as "writing" in some documentation.

Developing

The surface with the latent image is exposed to toner which has been applied in a 15-micron-thick layer to the developer roll. Toner consists of fine particles of dry plastic powder mixed with carbon black or coloring agents. The toner particles are given a negative charge inside the toner cartridge, and as they emerge onto the developer drum they are electrostatically attracted to the photoreceptor's latent image (the areas on the surface of the drum which had been struck by the laser). Because negative charges repel each other, the negatively charged toner particles will not adhere to the drum where the negative charge (imparted previously by the charge roller) remains.

Transferring

A sheet of paper is then rolled under the photoreceptor drum, which has been coated with a pattern of toner particles in the exact places where the laser struck it moments before. The toner particles have a very weak attraction to both the drum and the paper, but the bond to the drum is weaker and the particles transfer once again, this time from the drum's surface to the paper's surface. Some machines also use a positively charged "transfer roller" on the back side of the paper to help pull the negatively charged toner from the photoreceptor drum to the paper.

Fusing

The paper passes through rollers in the fuser assembly, where temperatures up to

200 °C (392 °F) and pressure are used to permanently bond the plastic powder to the paper. One roller is usually a hollow tube (heat roller) and the other is a rubber backed roller (pressure roller). A radiant heat lamp is suspended in the centre of the hollow tube, and its infrared energy uniformly heats the roller from the inside. For proper bonding of the toner, the fuser roller must be uniformly hot.

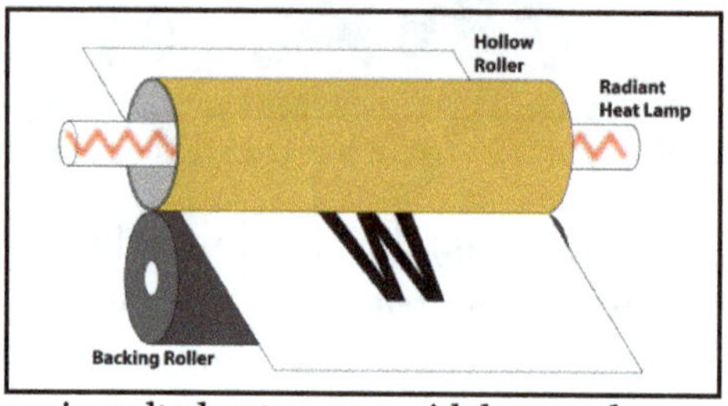

Toner is melted onto paper with heat and pressure.

Some printers use a very thin flexible metal foil roller, so there is less thermal mass to be heated and the fuser can more quickly reach operating temperature. If paper moves through the fuser more slowly, there is more roller contact time for the toner to melt, and the fuser can operate at a lower temperature. Smaller, inexpensive laser printers typically print slowly, due to this energy-saving design, compared to large high speed printers where paper moves more rapidly through a high-temperature fuser with a very short contact time.

Cleaning and Recharging

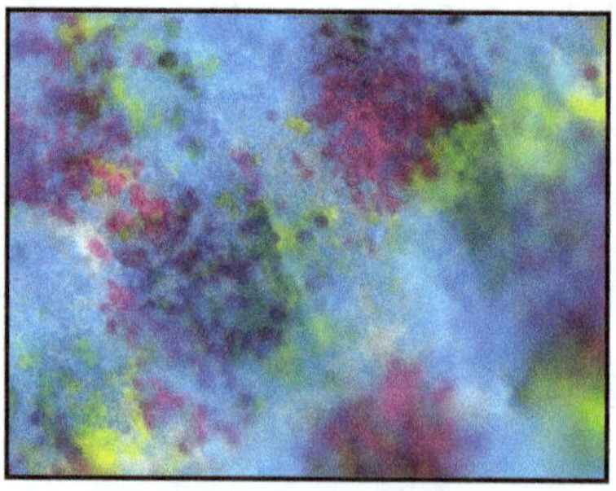

Magnification of color laser printer output, showing individual toner particles comprising 4 dots of an image with a bluish background.

As the drum completes a revolution, it is exposed to an electrically neutral soft plastic blade which cleans any remaining toner from the photoreceptor drum and deposits it into a waste reservoir. A charge roller then re-establishes a uniform negative charge on the surface of the now clean drum, readying it to be struck again by the laser.

Continuous Printing

Once the raster image generation is complete, all steps of the printing process can occur one after the other in rapid succession. This permits the use of a very small and compact unit, where the photoreceptor is charged, rotates a few degrees and is scanned, rotates a few more degrees and is developed, and so forth. The entire process can be completed before the drum completes one revolution.

Different printers implement these steps in distinct ways. LED printers actually use a linear array of light-emitting diodes to "write" the light on the drum. The toner is based on either wax or plastic, so that when the paper passes through the fuser assembly, the particles of toner melt. The paper may or may not be oppositely charged. The fuser can be an infrared oven, a heated pressure roller, or (on some very fast, expensive printers) a xenon flash lamp. The warmup process that a laser printer goes through when power is initially applied to the printer consists mainly of heating the fuser element.

Malfunctions

The mechanism inside a laser printer is somewhat delicate and, once damaged, often impossible to repair. The drum in particular is a critical component: it must not be left exposed to ambient light for more than a few hours, as light is what causes it to lose its charge and will eventually wear it out. Anything that interferes with the operation of the laser such as a scrap of torn paper may prevent the laser from discharging some portion of the drum, causing those areas to appear as white vertical streaks. If the neutral wiper blade fails to remove residual toner from the drum's surface, that toner may circulate on the drum a second time, causing smears on the printed page with each revolution. If the charge roller becomes damaged or does not have enough power, it may fail to adequately negatively charge the surface of the drum, allowing the drum to pick up excessive toner on the next revolution from the developer roll and causing a repeated but fainter image from the previous revolution to appear down the page.

If the toner doctor blade does not ensure that a smooth, even layer of toner is applied to the developer roll, the resulting printout may have white streaks from this in places where the blade has scraped off too much toner. Alternatively if the blade allows too much toner to remain on the developer roll, the toner particles might come loose as the roll turns, precipitate onto the paper below, and become bonded to the paper during the fusing process. This will result in a general darkening of the printed page in broad vertical stripes with very soft edges.

If the fuser roller does not reach a high enough temperature or if the ambient humidity is too high, the toner will not fuse well to the paper and may flake off after printing. If the fuser is too hot, the plastic component of the toner may smear, causing the printed text to look like it is wet or smudged, or may cause the melted toner to soak through the paper to the back side.

Different manufacturers claim that their toners are specifically developed for their printers, and that other toner formulations may not match the original specifications in terms of either tendency to accept a negative charge, to move to the discharged areas of the photoreceptor drum from the developer roll, to fuse appropriately to the paper, or to come off the drum cleanly in each revolution.

Performance

As with most electronic devices, the cost of laser printers has fallen markedly over the years. In 1984, the HP LaserJet sold for $3500, had trouble with even small, low resolution graphics, and weighed 32 kg (71 lb). As of 2016[update], low-end monochrome laser printers can sell for less than $75. These printers tend to lack onboard processing and rely on the host computer to generate a raster image, but outperform the 1984 LaserJet in nearly all situations.

Laser printer speed can vary widely, and depends on many factors, including the graphic intensity of the job being processed. The fastest models can print over 200 monochrome pages per minute (12,000 pages per hour). The fastest color laser printers can print over 100 pages per minute (6000 pages per hour). Very high-speed laser printers are used for mass mailings of personalized documents, such as credit card or utility bills, and are competing with lithography in some commercial applications.

The cost of this technology depends on a combination of factors, including the cost of paper, toner, drum replacement, as well as the replacement of other items such as the fuser assembly and transfer assembly. Often printers with soft plastic drums can have a very high cost of ownership that does not become apparent until the drum requires replacement.

Duplex printing (printing on both sides of the paper) can halve paper costs and reduce filing volumes. Formerly only available on high-end printers, duplexers are now common on mid-range office printers, though not all printers can accommodate a duplexing unit. Duplexing can also give a slower page-printing speed, because of the longer paper path.

Color Laser Printers

Fuji Xerox color laser printer C1110B.

Color laser printers use colored toner (dry ink), typically cyan, magenta, yellow, and black (CMYK). While monochrome printers only use one laser scanner assembly, color printers often have two or more.

Color printing adds complexity to the printing process because very slight misalignments known as registration errors can occur between printing each color, causing unintended color fringing, blurring, or light/dark streaking along the edges of colored regions. To permit a high registration accuracy, some color laser printers use a large rotating belt called a "transfer belt". The transfer belt passes in front of all the toner cartridges and each of the toner layers are precisely applied to the belt. The combined layers are then applied to the paper in a uniform single step.

Color printers usually have a higher cost per page than monochrome printers (even if printing monochrome-only pages).

Business Model Comparison with Inkjet Printers

Manufacturers use a similar business model for both low-cost color laser printers and inkjet printers: the printers are sold cheaply while replacement toners and inks are relatively expensive. Color laser printers are much faster than inkjet printers and their running cost per page is usually slightly less. The print quality of color lasers is limited by their resolution (typically 600–1200 dpi) and their use of just four color toners. They often have trouble printing large areas of the same or subtle gradations of color. Inkjet printers designed for printing photos can produce much higher quality color images.

Anti-Counterfeiting Marks

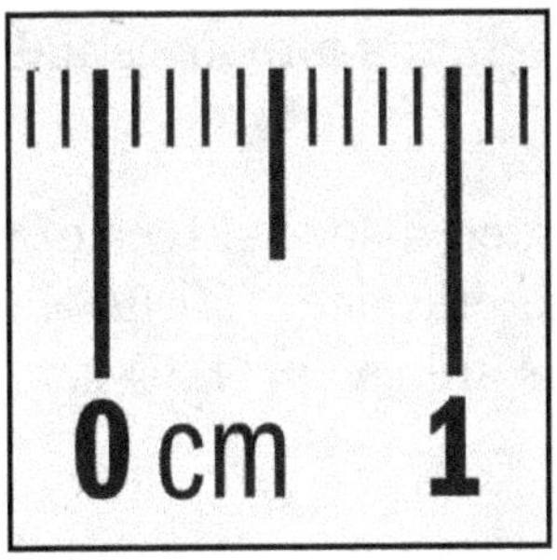

Small yellow dots on white paper, generated by a color laser printer, are nearly invisible.

Many modern color laser printers mark printouts by a nearly invisible dot raster, for the purpose of traceability. The dots are yellow and about 0.1 mm (0.0039 in) in size, with a raster of about 1 mm (0.039 in). This is purportedly the result of a deal between the US government and printer manufacturers to help track counterfeiters. The dots encode data such as printing date, time, and printer serial number in binary-coded decimal on every sheet of paper printed, which allows pieces of paper to be traced by the manufacturer to identify the place of purchase, and sometimes the buyer.

Digital rights advocacy groups such as the Electronic Frontier Foundation are concerned about this erosion of the privacy and anonymity of those who print.

Smart Chips in Toner Cartridges

Similar to inkjet printers, toner cartridges may contain smart chips that reduce the number of pages that can be printed with it (reducing the amount of usable ink in the cartridge to sometimes only 50%), in an effort to increase sales of the toner cartridges. Besides being more expensive to the consumer, this technique also increases waste, and thus increases pressure on the environment. For these toner cartridges (as with inkjet cartridges), reset devices can be used to override the limitation set by the smart chip. Also, for some printers, online walk-throughs have been posted to demonstrate how to use up all the ink in the cartridge. These chips offer no benefit to the end consumer— all laser printers originally used an optical mechanism to assess the amount of remaining toner in the cartridge rather than using a chip to electrically count the number of printed pages, and the chip's only function was as an alternate method to decrease the cartridge's usable life.

Safety Hazards, Health Risks, and Precautions

Toner Clean-Up

Toner particles are formulated to have electrostatic properties and can develop static electric charges when they rub against other particles, objects, or the interiors of transport systems and vacuum hoses. Static discharge from charged toner particles can ignite combustible particles in a vacuum cleaner bag or create a small dust explosion if sufficient toner is airborne. Toner particles are so fine that they are poorly filtered by conventional household vacuum cleaner filter bags and blow through the motor or back into the room.

If toner spills into the laser printer, a special type of vacuum cleaner with an electrically conductive hose and a high-efficiency (HEPA) filter may be needed for effective cleaning. These specialized tools are called "ESD-safe" (Electrostatic Discharge-safe) or "toner vacuums".

Ozone Hazards

As a normal part of the printing process, the high voltages inside the printer can produce a corona discharge that generates a small amount of ionized oxygen and nitrogen, which react to form ozone and nitrogen oxides. In larger commercial printers and copiers, an activated carbon filter in the air exhaust stream breaks down these noxious gases to prevent pollution of the office environment.

However, some ozone escapes the filtering process in commercial printers, and ozone filters are not used at all in most smaller consumer printers. When a laser printer or copier is operated for a long period of time in a small, poorly ventilated space, these gases can build up to levels at which the odor of ozone or irritation may be noticed. A potential for creating a health hazard is theoretically possible in extreme cases.

Respiratory Health Risks

According to a 2012 study conducted in Queensland, Australia, some printers emit sub-micrometre particles which some suspect may be associated with respiratory diseases. Of 63 printers evaluated in the Queensland University of Technology study, 17 of the strongest emitters were made by HP and one by Toshiba. The machine population studied, however, was only those machines already in place in the building and was thus biased toward specific manufacturers. The authors noted that particle emissions varied substantially even among the same model of machine. According to Professor Morawska of Queensland University, one printer emitted as many particles as a burning cigarette:

The health effects from inhaling ultrafine particles depend on particle composition, but the results can range from respiratory irritation to more severe illness such as cardiovascular problems or cancer.

— *Queensland University of Technology*

Muhle et al. (1991) reported that the responses to chronically inhaled copying toner, a plastic dust pigmented with carbon black, titanium dioxide and silica were also similar qualitatively to titanium dioxide and diesel exhaust.

In December 2011, the Australian government agency Safe Work Australia reviewed existing research and concluded that "no epidemiology studies directly associating laser printer emissions with adverse health outcomes were located" and that several assessments conclude that "risk of direct toxicity and health effects from exposure to laser printer emissions is negligible". The review also observes that, because the emissions have been shown to be volatile or semi-volatile organic compounds, "it would be logical to expect possible health effects to be more related to the chemical nature of the aerosol rather than the physical character of the 'particulate' since such emissions are unlikely to be or remain as 'particulates' after they come into contact with respiratory tissue".

Air Transport Ban

After the 2010 cargo plane bomb plot, in which shipments of laser printers with explosive-filled toner cartridges were discovered on separate cargo airplanes, the US Transportation Security Administration prohibited pass-through passengers from carrying toner or ink cartridges weighing over 1 pound (0.45 kg) on inbound flights, in both carry-on and checked luggage. *PC Magazine noted that the ban would not impact most travelers, as the majority of cartridges do not exceed the proscribed weight.*

3D Scanner

A 3D scanner is a device that analyses a real-world object or environment to collect data on its shape and possibly its appearance (e.g. colour). The collected data can then be used to construct digital three-dimensional models.

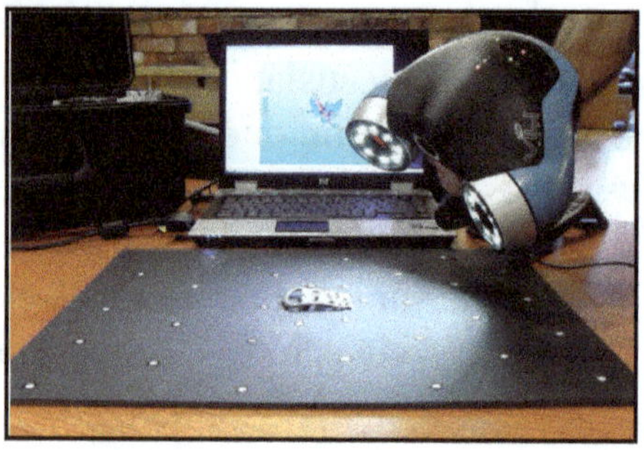

Three-dimensional scanner used to create 3D animation and special effects.

Many different technologies can be used to build these 3D-scanning devices; each technology comes with its own limitations, advantages and costs. Many limitations in the kind of objects that can be digitised are still present, for example, optical technologies encounter many difficulties with shiny, mirroring or transparent objects. For example, industrial computed tomography scanning can be used to construct digital 3D models, applying non-destructive testing.

3D scanned interior of St Joseph's Church, Subiaco.

Collected 3D data is useful for a wide variety of applications. These devices are used extensively by the entertainment industry in the production of movies and video games. Other common applications of this technology include industrial design, orthotics and prosthetics, reverse engineering and prototyping, quality control/inspection and documentation of cultural artifacts.t

Functionality

The purpose of a 3D scanner is usually to create a point cloud of geometric samples on the surface of the subject. These points can then be used to extrapolate the shape of the subject (a process called reconstruction). If colour information is collected at each point, then the colours on the surface of the subject can also be determined.

3D scanning of a fin whale skeleton in the Natural History Museum of Slovenia (August 2013)

3D scanners share several traits with cameras. Like most cameras, they have a cone-like field of view, and like cameras, they can only collect information about surfaces that are not obscured. While a camera collects colour information about surfaces within its field of view, a 3D scanner collects distance information about surfaces within its field of view. The "picture" produced by a 3D scanner describes the distance to a surface at each point in the picture. This allows the three dimensional position of each point in the picture to be identified.

For most situations, a single scan will not produce a complete model of the subject. Multiple scans, even hundreds, from many different directions are usually required to obtain information about all sides of the subject. These scans have to be brought into a common reference system, a process that is usually called *alignment or registration, and then merged to create a complete model. This whole process, going from the single range map to the whole model, is usually known as the 3D scanning pipeline.*

Technology

There are a variety of technologies for digitally acquiring the shape of a 3D object. A well established classification divides them into two types: contact and non-contact. Non-contact solutions can be further divided into two main categories, active and passive. There are a variety of technologies that fall under each of these categories.

Contact

Contact 3D scanners probe the subject through physical touch, while the object is in contact with or resting on a precision flat surface plate, ground and polished to a specific maximum of surface roughness. Where the object to be scanned is not flat or can not rest stably on a flat surface, it is supported and held firmly in place by a fixture.

The scanner mechanism may have three different forms:

A carriage system with rigid arms held tightly in perpendicular relationship and each axis gliding along a track. Such systems work best with flat profile shapes or simple convex curved surfaces.

A coordinate measuring machine with rigid perpendicular arms.

An articulated arm with rigid bones and high precision angular sensors. The location of the end of the arm involves complex math calculating the wrist rotation angle and hinge angle of each joint. This is ideal for probing into crevasses and interior spaces with a small mouth opening.

A combination of both methods may be used, such as an articulated arm suspended from a traveling carriage, for mapping large objects with interior cavities or overlapping surfaces.

A CMM (coordinate measuring machine) is an example of a contact 3D scanner. It is used mostly in manufacturing and can be very precise. The disadvantage of CMMs though, is that it requires contact with the object being scanned. Thus, the act of scanning the object might modify or damage it. This fact is very significant when scanning delicate or valuable objects such as historical artifacts. The other disadvantage of CMMs is that they are relatively slow compared to the other scanning methods. Physically moving the arm that the probe is mounted on can be very slow and the fastest CMMs can only operate on a few hundred hertz. In contrast, an optical system like a laser scanner can operate from 10 to 500 kHz.

Other examples are the hand driven touch probes used to digitise clay models in computer animation industry.

Non-Contact Active

Active scanners emit some kind of radiation or light and detect its reflection or radiation passing through object in order to probe an object or environment. Possible types of emissions used include light, ultrasound or x-ray.

Time-Of-Flight

The time-of-flight 3D laser scanner is an active scanner that uses laser light to probe the subject. At the heart of this type of scanner is a time-of-flight laser range finder.

The laser range finder finds the distance of a surface by timing the round-trip time of a pulse of light. A laser is used to emit a pulse of light and the amount of time before the reflected light is seen by a detector is measured. Since the speed of light *c* is known, the round-trip time determines the travel distance of the light, which is twice the distance between the scanner and the surface. If t is the round-trip time, then distance is equal to $c \cdot t / 2$. The accuracy of a time-of-flight 3D laser scanner depends on how precisely we can measure the *t* time: 3.3 picoseconds (approx.) is the time taken for light to travel 1 millimetre.

This lidar scanner may be used to scan buildings, rock formations, etc., to produce a 3D model. The lidar can aim its laser beam in a wide range: its head rotates horizontally, a mirror flips vertically. The laser beam is used to measure the distance to the first object on its path.

The laser range finder only detects the distance of one point in its direction of view. Thus, the scanner scans its entire field of view one point at a time by changing the range finder's direction of view to scan different points. The view direction of the laser range finder can be changed either by rotating the range finder itself, or by using a system of rotating mirrors. The latter method is commonly used because mirrors are much lighter and can thus be rotated much faster and with greater accuracy. Typical time-of-flight 3D laser scanners can measure the distance of 10,000~100,000 points every second.

Time-of-flight devices are also available in a 2D configuration. This is referred to as a time-of-flight camera.

Triangulation

Triangulation based 3D laser scanners are also active scanners that use laser light to probe the environment. With respect to time-of-flight 3D laser scanner the triangulation laser shines a laser on the subject and exploits a camera to look for the location of the laser dot. Depending on how far away the laser strikes a surface, the laser dot appears at different places in the camera's field of view. This technique is called triangulation because the laser dot, the camera and the laser emitter form a triangle. The length of one side of the triangle, the distance between the camera and the laser emitter is known. The angle of the laser emitter corner is also known. The angle of the camera corner can be determined by looking at the location of the laser dot in the camera's field of view. These three pieces of information fully determine the shape and size of the triangle and give the location of the laser dot corner of the triangle. In most cases a laser stripe, instead of a single laser dot, is swept across the object to speed up the acquisition process. The National Research Council of Canada was among the first institutes to develop the triangulation based laser scanning technology in 1978.

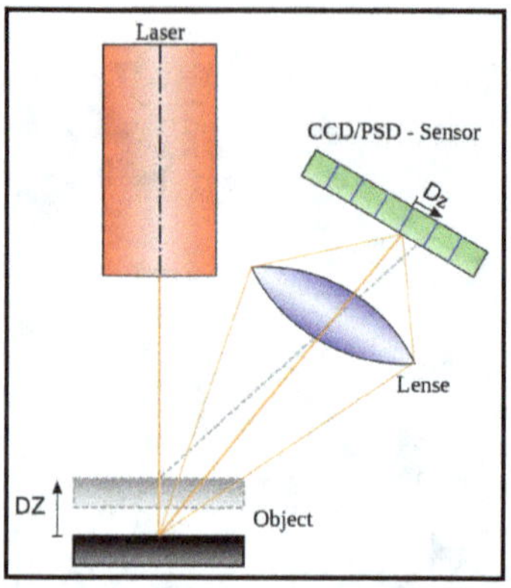

Principle of a laser triangulation sensor. Two object positions are shown.

Strengths and Weaknesses

Time-of-flight and triangulation range finders each have strengths and weaknesses that make them suitable for different situations. The advantage of time-of-flight range finders is that they are capable of operating over very long distances, on the order of kilometres. These scanners are thus suitable for scanning large structures like buildings or geographic features. The disadvantage of time-of-flight range finders is their accuracy. Due to the high speed of light, timing the round-trip time is difficult and the accuracy of the distance measurement is relatively low, on the order of millimetres.

Triangulation range finders are exactly the opposite. They have a limited range of some meters, but their accuracy is relatively high. The accuracy of triangulation range finders is on the order of tens of micrometers.

Time-of-flight scanners' accuracy can be lost when the laser hits the edge of an object because the information that is sent back to the scanner is from two different locations for one laser pulse. The coordinate relative to the scanner's position for a point that has hit the edge of an object will be calculated based on an average and therefore will put the point in the wrong place. When using a high resolution scan on an object the chances of the beam hitting an edge are increased and the resulting data will show noise just behind the edges of the object. Scanners with a smaller beam width will help to solve this problem but will be limited by range as the beam width will increase over distance. Software can also help by determining that the first object to be hit by the laser beam should cancel out the second.

At a rate of 10,000 sample points per second, low resolution scans can take less than a second, but high resolution scans, requiring millions of samples, can take minutes for some time-of-flight scanners. The problem this creates is distortion from motion. Since each point is sampled at a different time, any motion in the subject or the scanner will distort the collected data. Thus, it is usually necessary to mount both the subject and the scanner on stable platforms and minimise vibration. Using these scanners to scan objects in motion is very difficult.

Recently, there has been research on compensating for distortion from small amounts of vibration and distortions due to motion and/or rotation.

When scanning in one position for any length of time slight movement can occur in the scanner position due to changes in temperature. If the scanner is set on a tripod and there is strong sunlight on one side of the scanner then that side of the tripod will expand and slowly distort the scan data from one side to another. Some laser scanners have level compensators built into them to counteract any movement of the scanner during the scan process.

In a conoscopic system, a laser beam is projected onto the surface and then the immediate reflection along the same ray-path are put through a conoscopic crystal and projected onto a CCD. The result is a diffraction pattern, that can be frequency analyzed to determine the distance to the measured surface. The main advantage with conoscopic holography is that only a single ray-path is needed for measuring, thus giving an opportunity to measure for instance the depth of a finely drilled hole.

Using a periscope allows to into small diameter holes and measure bottom and side walls.

Hand-Held Laser Scanners

Hand-held laser scanners create a 3D image through the triangulation mechanism described above: a laser dot or line is projected onto an object from a hand-held device and a sensor (typically a charge-coupled device or position sensitive device) measures the distance to the surface. Data is collected in relation to an internal coordinate system and therefore to collect data where the scanner is in motion the position of the scanner must be determined. The position can be determined by the scanner using reference features on the surface being scanned (typically adhesive reflective tabs, but natural features have been also used in research work) or by using an external tracking method. External tracking often takes the form of a laser tracker (to provide the sensor position) with integrated camera (to determine the orientation of the scanner) or a photogrammetric solution using 3 or more cameras providing the complete Six degrees of freedom of the scanner. Both techniques tend to use infra red Light-emitting diodes attached to the scanner which are seen by the camera(s) through filters providing resilience to ambient lighting.

Data is collected by a computer and recorded as data points within Three-dimensional space, with processing this can be converted into a triangulated mesh and then a Computer-aided design model, often as Non-uniform rational B-spline surfaces. Hand-held laser scanners can combine this data with passive, visible-light sensors — which capture surface textures and colours — to build (or "reverse engineer") a full 3D model.

Structured Light

Structured-light 3D scanners project a pattern of light on the subject and look at the deformation of the pattern on the subject. The pattern is projected onto the subject using either an LCD projector or other stable light source. A camera, offset slightly from the pattern projector, looks at the shape of the pattern and calculates the distance of every point in the field of view.

Structured-light scanning is still a very active area of research with many research papers published each year. Perfect maps have also been proven useful as structured light patterns that solve the correspondence problem and allow for error detection and error correction.

The advantage of structured-light 3D scanners is speed and precision. Instead of scanning one point at a time, structured light scanners scan multiple points or the entire field of view at once. Scanning an entire field of view in a fraction of a second reduces or eliminates the problem of distortion from motion. Some existing systems are capable of scanning moving objects in real-time. VisionMaster creates a 3D scanning system with a 5-megapixel camera – 5 million data points are acquired in every frame.

A real-time scanner using digital fringe projection and phase-shifting technique (certain kinds of structured light methods) was developed, to capture, reconstruct, and render high-density details of dynamically deformable objects (such as facial expressions) at 40 frames per second. Recently, another scanner has been developed. Different patterns can be applied to this system, and the frame rate for capturing and data processing achieves 120 frames per second. It can also scan isolated surfaces, for example two moving hands. By utilising the binary defocusing technique, speed breakthroughs have been made that could reach hundreds of to thousands of frames per second.

Modulated Light

Modulated light 3D scanners shine a continually changing light at the subject. Usually the light source simply cycles its amplitude in a sinusoidal pattern. A camera detects the reflected light and the amount the pattern is shifted by determines the distance the light travelled. Modulated light also allows the scanner to ignore light from sources other than a laser, so there is no interference.

Volumetric Techniques

Medical

Computed tomography (CT) is a medical imaging method which generates a three-dimensional image of the inside of an object from a large series of two-dimensional X-ray images, similarly Magnetic resonance imaging is another medical imaging technique that provides much greater contrast between the different soft tissues of the body than computed tomography (CT) does, making it especially useful in neurological (brain), musculoskeletal, cardiovascular, and oncological (cancer) imaging. These techniques produce a discrete 3D volumetric representation that can be directly visualised, manipulated or converted to traditional 3D surface by mean of isosurface extraction algorithms.

Industrial

Although most common in medicine, Industrial computed tomography, Microtomography and MRI are also used in other fields for acquiring a digital representation of an object and its interior, such as non destructive materials testing, reverse engineering, or studying biological and paleontological specimens.

Non-Contact Passive

Passive 3D imaging solutions do not emit any kind of radiation themselves, but instead rely on detecting reflected ambient radiation. Most solutions of this type detect visible light because it is a readily available ambient radiation. Other types of radiation, such as infra red could also be used. Passive methods can be very cheap, because in most cases they do not need particular hardware but simple digital cameras.

- *Stereoscopic systems usually employ two video cameras, slightly apart, looking at the same scene. By analysing the slight differences between the images seen by each camera, it is possible to determine the distance at each point in the images. This method is based on the same principles driving human* stereoscopic vision.

- *Photometric systems usually use a single camera, but take multiple images under varying lighting conditions. These techniques attempt to invert the image formation model in order to recover the surface orientation at each pixel.*
- *Silhouette techniques use outlines created from a sequence of photographs around a three-dimensional object against a well contrasted background. These* silhouettes are extruded and intersected to form the visual hull approximation of the object. With these approaches some concavities of an object (like the interior of a bowl) cannot be detected.

User Assisted (Image-Based Modelling)

There are other methods that, based on the user assisted detection and identification of some features and shapes on a set of different pictures of an object are able to build an approximation of the object itself. This kind of techniques are useful to build fast approximation of simple shaped objects like buildings. Various commercial packages are available like D-Sculptor, iModeller, Autodesk ImageModeler, 123DCatch or PhotoModeler.

This sort of 3D imaging solution is based on the principles of photogrammetry. It is also somewhat similar in methodology to panoramic photography, except that the photos are taken of one object on a three-dimensional space in order to replicate it instead of taking a series of photos from one point in a three-dimensional space in order to replicate the surrounding environment.

Reconstruction

From Point Clouds

The point clouds produced by 3D scanners and 3D imaging can be used directly for measurement and visualisation in the architecture and construction world.

From Models

Most applications, however, use instead polygonal 3D models, NURBS surface models, or editable feature-based CAD models (aka Solid models).

- Polygon mesh models: In a polygonal representation of a shape, a curved surface is modeled as many small faceted flat surfaces (think of a sphere modeled as a disco ball). Polygon models—also called Mesh models, are useful for visualisation, for some CAM (i.e., machining), but are generally “heavy” (i.e., very large data sets), and are relatively un-editable in this form. Reconstruction to polygonal model involves finding and connecting adjacent points with straight lines in order to create a continuous surface. Many applications, both free and nonfree, are available for this purpose (e.g. MeshLab, PointCab, kubit PointCloud for AutoCAD, JRC 3D Reconstructor, imagemodel, PolyWorks, Rapidform, Geomagic, Imageware, Rhino 3D etc.).

- Surface models: The next level of sophistication in modeling involves using a quilt of *curved surface patches to model our shape. These might be NURBS, TSplines or other curved representations of curved topology. Using NURBS, our sphere is a true mathematical sphere. Some applications offer patch layout by hand but the best in class offer both automated patch layout and manual layout. These patches have the advantage of being lighter and more manipulable when exported to CAD. Surface models are somewhat editable, but only in a sculptural sense of pushing and pulling to deform the surface. This representation lends itself well to modelling organic and artistic shapes. Providers of surface modellers include Rapidform,* Geomagic, Rhino 3D, Maya, T Splines etc.
- Solid CAD models: From an engineering/manufacturing perspective, the ultimate representation of a digitised shape is the editable, parametric CAD model. After all, CAD is the common "language" of industry to describe, edit and maintain the shape of the enterprise's assets. In CAD, our sphere is described by parametric features which are easily edited by changing a value (e.g., centre point and radius).

These CAD models describe not simply the envelope or shape of the object, but CAD models also embody the "design intent" (i.e., critical features and their relationship to other features). An example of design intent not evident in the shape alone might be a brake drum's lug bolts, which must be concentric with the hole in the centre of the drum. This knowledge would drive the sequence and method of creating the CAD model; a designer with an awareness of this relationship would not design the lug bolts referenced to the outside diameter, but instead, to the center. A modeler creating a CAD model will want to include both Shape and design intent in the complete CAD model.

Vendors offer different approaches to getting to the parametric CAD model. Some export the NURBS surfaces and leave it to the CAD designer to complete the model in CAD (e.g., Geomagic, Imageware, Rhino 3D). Others use the scan data to create an editable and verifiable feature based model that is imported into CAD with full feature tree intact, yielding a complete, native CAD model, capturing both shape and design intent (e.g. Geomagic, Rapidform). Still other CAD applications are robust enough to manipulate limited points or polygon models within the CAD environment (e.g., CATIA, AutoCAD, Revit).

From a Set of 2D Slices

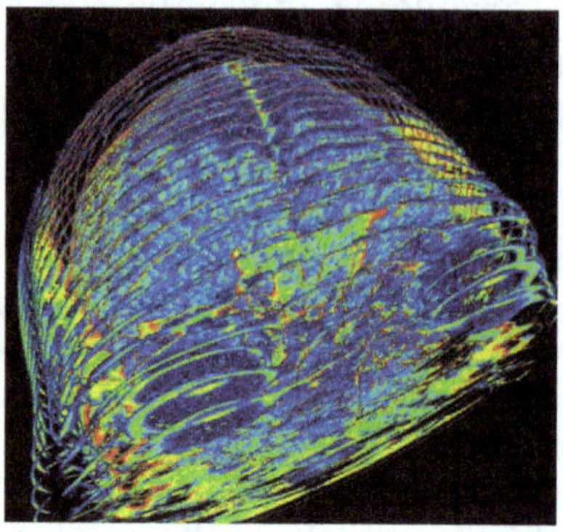

3D reconstruction of the brain and eyeballs from CT scanned DICOM images. In this image, areas with the density of bone or air were made transparent, and the slices stacked up in an approximate free-space alignment. The outer ring of material around the brain are the soft tissues of skin and muscle on the outside of the skull. A black box encloses the slices to provide the black background. Since these are simply 2D images stacked up, when viewed on edge the slices disappear since they have effectively zero thickness. Each DICOM scan represents about 5mm of material averaged into a thin slice.

CT, industrial CT, MRI, or Micro-CT scanners do not produce point clouds but a set of 2D slices (each termed a "tomogram") which are then 'stacked together' to produce a 3D representation. There are several ways to do this depending on the output required:

- Volume rendering: Different parts of an object usually have different threshold values or greyscale densities. From this, a 3-dimensional model can be constructed and displayed on screen. Multiple models can be constructed from various thresholds, allowing different colours to represent each component of the object. Volume rendering is usually only used for visualisation of the scanned object.
- Image segmentation: Where different structures have similar threshold/greyscale values, it can become impossible to separate them simply by adjusting volume rendering parameters. The solution is called segmentation, a manual or automatic procedure that can remove the unwanted structures from the image. Image segmentation software usually allows export of the segmented structures in CAD or STL format for further manipulation.
- Image-based meshing: When using 3D image data for computational analysis (e.g. CFD and FEA), simply segmenting the data and meshing from CAD can become time consuming, and virtually intractable for the complex topologies typical of image data. The solution is called image-based meshing, an automated process of generating an accurate and realistic geometrical description of the scan data.

From Laser Scans

Laser scanning describes the general method to sample or scan a surface using laser technology. Several areas of application exist that mainly differ in the power of the lasers that are used, and in the results of the scanning process. Low laser power is used when the scanned surface doesn't have to be influenced, e.g. when it only has to be digitised. Confocal or 3D laser scanning are methods to get information about the scanned surface. Another low-power application uses structured light projection systems for solar cell flatness metrology, enabling stress calculation throughout in excess of 2000 wafers per hour.

The laser power used for laser scanning equipment in industrial applications is typically less than 1W. The power level is usually on the order of 200 mW or less but sometime more.

Applications

Construction industry and Civil Engineering

- Robotic control: e.g. a laser scanner may function as the "eye" of a robot.
- As-built drawings of bridges, industrial plants, and monuments.
- Documentation of historical sites.
- Site modelling and lay outing.
- Quality control.
- Quantity surveys.
- Freeway redesign.
- Establishing a bench mark of pre-existing shape/state in order to detect structural changes resulting from exposure to extreme loadings such as earthquake, vessel/truck impact or fire.
- Create GIS (geographic information system) maps and geomatics.
- Subsurface laser scanning in mines and Karst voids.
- Forensic documentation.

Design Process

- Increasing accuracy working with complex parts and shapes.
- Coordinating product design using parts from multiple sources.
- Updating old CD scans with those from more current technology.
- Replacing missing or older parts.
- Creating cost savings by allowing as-built design services, for example in automotive manufacturing plants.

- "Bringing the plant to the engineers" with web shared scans.
- Saving travel costs.

Entertainment

3D scanners are used by the entertainment industry to create digital 3D models for movies, video games and leisure purposes. They are heavily utilised in virtual cinematography. In cases where a real-world equivalent of a model exists, it is much faster to scan the real-world object than to manually create a model using 3D modeling software. Frequently, artists sculpt physical models of what they want and scan them into digital form rather than directly creating digital models on a computer.

Law Enforcment

3D laser scanning is used by the FBI. 3D Models are used for on-site documentation of:

- Crime scenes.
- Bullet trajectory.
- Accident reconstruction.
- Bombings.
- Plane crashes, and more.

Reverse Engineering

Reverse engineering of a mechanical component requires a precise digital model of the objects to be reproduced. Rather than a set of points a precise digital model can be represented by a polygon mesh, a set of flat or curved NURBS surfaces, or ideally for mechanical components, a CAD solid model. A 3D scanner can be used to digitise free-form or gradually changing shaped components as well as prismatic geometries whereas a coordinate measuring machine is usually used only to determine simple dimensions of a highly prismatic model. These data points are then processed to create a usable digital model, usually using specialized reverse engineering software.

Cultural Heritage

There have been many research projects undertaken via the scanning of historical sites and artifacts both for documentation and analysis purposes.

The combined use of 3D scanning and 3D printing technologies allows the replication of real objects without the use of traditional plaster casting techniques, that in many cases can be too invasive for being performed on precious or delicate cultural heritage artifacts. In an example of a typical application scenario, a gargoyle model was digitally acquired using a 3D scanner and the produced 3D data was processed using MeshLab. The resulting digital 3D model was fed to a rapid prototyping machine to create a real resin replica of the original object.

Michelangelo

In 1999, two different research groups started scanning Michelangelo's statues. Stanford University with a group led by Marc Levoy used a custom laser triangulation scanner built by Cyberware to scan Michelangelo's statues in Florence, notably the David, the Prigioni and the four statues in The Medici Chapel. The scans produced a data point density of one sample per 0.25 mm, detailed enough to see Michelangelo's chisel marks. These detailed scans produced a large amount of data (up to 32 gigabytes) and processing the data from his scans took 5 months. Approximately in the same period a research group from IBM, led by H. Rushmeier and F. Bernardini scanned the Pietà of Florence acquiring both geometric and colour details. The digital model, result of the Stanford scanning campaign, was thoroughly used in the 2004 subsequent restoration of the statue.

Monticello

In 2002, David Luebke, et al. scanned Thomas Jefferson's Monticello. A commercial time of flight laser scanner, the DeltaSphere 3000, was used. The scanner data was later combined with colour data from digital photographs to create the Virtual Monticello, and the Jefferson's Cabinet exhibits in the New Orleans Museum of Art in 2003. The Virtual Monticello exhibit simulated a window looking into Jefferson's Library. The exhibit consisted of a rear projection display on a wall and a pair of stereo glasses for the viewer. The glasses, combined with polarised projectors, provided a 3D effect. Position tracking hardware on the glasses allowed the display to adapt as the viewer moves around, creating the illusion that the display is actually a hole in the wall looking into Jefferson's Library. The Jefferson's Cabinet exhibit was a barrier stereogram (essentially a non-active hologram that appears different from different angles) of Jefferson's Cabinet.

Cuneiform Tablets

In 2003, Subodh Kumar, et al. undertook the 3D scanning of ancient cuneiform tablets. Again, a laser triangulation scanner was used. The tablets were scanned on a regular grid pattern at a resolution of 0.025 mm (0.00098 in).

Kasubi Tombs

A 2009 CyArk 3D scanning project at Uganda's historic Kasubi Tombs, a UNESCO World Heritage Site, using a Leica HDS 4500, produced detailed architectural models of Muzibu Azaala Mpanga, the main building at the complex and tomb of the Kabakas (Kings) of Uganda. A fire on March 16, 2010, burned down much of the Muzibu Azaala Mpanga structure, and reconstruction work is likely to lean heavily upon the dataset produced by the 3D scan mission.

"Plastico Di Roma Antica"

In 2005, Gabriele Guidi, et al. scanned the "Plastico di Roma antica", a model of Rome created in the last century. Neither the triangulation method, nor the time of flight method satisfied the requirements of this project because the item to be scanned was both large and contained small details. They found though, that a modulated light scanner was able to provide both the ability to scan an object the size of the model and the accuracy that was needed. The modulated light scanner was supplemented by a triangulation scanner which was used to scan some parts of the model.

Other Projects

The 3D Encounters Project at the Petrie Museum of Egyptian Archaeology aims to use 3D laser scanning to create a high quality 3D image library of artefacts and enable digital travelling exhibitions of fragile Egyptian artefacts, English Heritage has investigated the use of 3D laser scanning for a wide range of applications to gain archaeological and condition data, and the National Conservation Centre in Liverpool has also produced 3D laser scans on commission, including portable object and in situ scans of archaeological sites. The Smithsonian Institution has a project called Smithsonian X 3D notable for the breadth of types of 3D objects they are attempting to scan. These include small objects such as insects and flowers, to human sized objects such as Amelia Earhart's Flight Suit to room sized objects such as the Gunboat Philadelphia to historic sites such as Liang Bua in Indonesia. Also of note the data from these scans is being made available to the public for free and downloadable in several data formats.

Medical CAD/CAM

3D scanners are used to capture the 3D shape of a patient in orthotics and dentistry. It gradually supplants tedious plaster cast. CAD/CAM software are then used to design and manufacture the orthosis, prosthesis or dental implants.

Many Chairside dental CAD/CAM systems and Dental Laboratory CAD/CAM systems use 3D Scanner technologies to capture the 3D surface of a dental preparation (either *in vivo or in vitro), in order to produce a restoration digitally using CAD software and ultimately produce the final restoration using a CAM technology (such as a CNC milling machine, or 3D printer). The chairside systems are designed to facilitate the 3D scanning of a preparation in vivo and produce the restoration (such as a Crown, Onlay, Inlay or Veneer).*

Quality Assurance and Industrial Metrology

The digitalisation of real-world objects is of vital importance in various application domains. This method is especially applied in industrial quality assurance to measure the geometric dimension accuracy. Industrial processes such as assembly are complex, highly automated and typically based on CAD (Computer Aided Design) data. The problem is that the same degree of automation is also required for quality assurance. It is, for example, a very complex task to assemble a modern car, since it consists of many parts that must fit together at the very end of the production line. The optimal performance of this process is guaranteed by quality assurance systems. Especially the geometry of the metal parts must be checked in order to assure that they have the correct dimensions, fit together and finally work reliably.

Within highly automated processes, the resulting geometric measures are transferred to machines that manufacture the desired objects. Due to mechanical uncertainties and abrasions, the result may differ from its digital nominal. In order to automatically capture and evaluate these deviations, the manufactured part must be digitised as well. For this purpose, 3D scanners are applied to generate point samples from the object's surface which are finally compared against the nominal data.

The process of comparing 3D data against a CAD model is referred to as CAD-Compare, and can be a useful technique for applications such as determining wear patterns on moulds and tooling, determining accuracy of final build, analysing gap and flush, or analysing highly complex sculpted surfaces. At present, laser triangulation scanners, structured light and contact scanning are the predominant technologies employed for industrial purposes, with contact scanning remaining the slowest, but overall most accurate option.

Barcode Reader

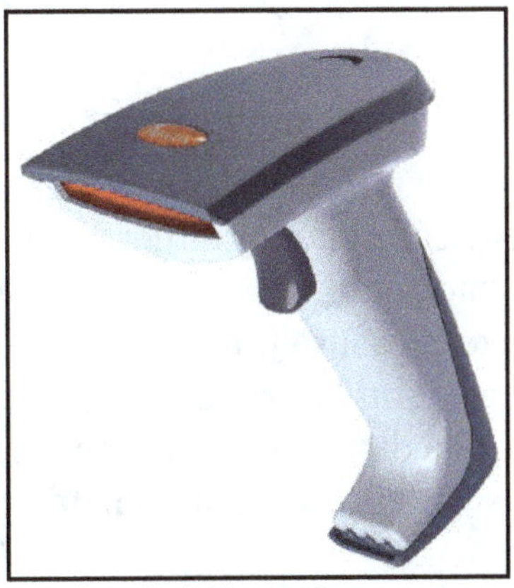

A handheld barcode scanner.

A barcode reader (or barcode scanner) is an electronic device that can read and output printed barcodes to a computer. Like a flatbed scanner, it consists of a light source, a lens and a light sensor translating optical impulses into electrical ones. Additionally, nearly all barcode readers contain *decoder circuitry analyzing the barcode's image data provided by the sensor and sending the barcode's content to the scanner's output port.*

A stationary barcode scanner for a conveyor line.

Types of Barcode Scanners

Technology

Barcode readers can be differentiated by technologies as follows:

Pen-type scanners Pen-type readers consist of a light source and photodiode that are placed next to each other in the tip of a pen or wand. To read a bar code, the person holding the pen must move the tip of it across the bars at a relatively uniform speed. The photodiode measures the intensity of the light reflected back from the light source as the tip crosses each bar and space in the printed code. The photodiode generates a waveform that is used to measure the widths of the bars and spaces in the bar code. Dark bars in the bar code absorb light and white spaces reflect light so that the voltage waveform generated by the photodiode is a representation of the bar and space pattern in the bar code. This waveform is decoded by the scanner in a manner similar to the way Morse code dots and dashes are decoded.

Laser Scanners

Laser scanners work the same way as pen type readers except that they use a laser beam as the light source and typically employ either a reciprocating mirror or a rotating prism to scan the laser beam back and forth across the bar code. As with the pen type reader, a photo-diode is used to measure the intensity of the light reflected back from the bar code. In both pen readers and laser scanners, the light emitted by the reader is rapidly varied in brightness with a data pattern and the photo-diode receive circuitry is designed to detect only signals with the same modulated pattern.

CCD readers (Also Known as LED Scanners)

CCD readers use an array of hundreds of tiny light sensors lined up in a row in the head of the reader. Each sensor measures the intensity of the light immediately in front of it. Each individual light sensor in the CCD reader is extremely small and because there are hundreds of sensors lined up in a row, a voltage pattern identical to the pattern in a bar code is generated in the reader by sequentially measuring the voltages across each sensor in the row. The important difference between a CCD reader and a pen or laser scanner is that the CCD reader is measuring emitted ambient light from the bar code whereas pen or laser scanners are measuring reflected light of a specific frequency originating from the scanner itself.

Camera-Based Readers

Two-dimensional imaging scanners are a newer type of bar code reader. They use a camera and image processing techniques to decode the bar code.

vVideo camera readers use small video cameras with the same CCD technology as in a CCD bar code reader except that instead of having a single row of sensors, a video camera has hundreds of rows of sensors arranged in a two dimensional array so that they can generate an image.

Large field-of-view readers use high resolution industrial cameras to capture multiple bar codes simultaneously. All the bar codes appearing in the photo are decoded instantly (ImageID patents and code creation tools) or by use of plugins (e.g. the Barcodepedia used a flash application and some web cam for querying a database), have been realized options for resolving the given tasks.

Omnidirectional Barcode Scanners

Omnidirectional scanning uses "series of straight or curved scanning lines of varying directions in the form of a starburst, a Lissajous pattern, or other multiangle arrangement are projected at the symbol and one or more of them will be able to cross all of the symbol's bars and spaces, no matter what the orientation."

Omnidirectional scanners almost all use a laser. Unlike the simpler single-line laser scanners, they produce a pattern of beams in varying orientations allowing them to read barcodes presented to it at different angles. Most of them use a single rotating polygonal mirror and an arrangement of several fixed mirrors to generate their complex scan patterns.

Omnidirectional scanners are most familiar through the horizontal scanners in supermarkets, where packages are slid over a glass or sapphire window. There are a range of different omnidirectional units available which can be used for differing scanning applications, ranging from retail type applications with the barcodes read only a few centimetres away from the scanner to industrial conveyor scanning where the unit can be a couple of metres away or more from the code. Omnidirectional scanners are also better at reading poorly printed, wrinkled, or even torn barcodes.

Cell Phone Cameras

While cell phone cameras without auto-focus are not ideal for reading some common barcode formats, there are 2D barcodes which are optimized for cell phones, as well as QR Codes and Data Matrix codes which can be read quickly and accurately with or without auto-focus.

Cell phone cameras open up a number of applications for consumers:

- Movies: DVD/VHS movie catalogs.
- Music: CD catalogs – play MP3 when scanned.
- Book catalogs and device.
- Groceries, nutrition information, making shopping lists when the last of an item is used, etc.
- Personal Property inventory (for insurance and other purposes)ode scanned into personal finance software when entering. Later, scanned receipt images can then be automatically associated with the appropriate entries. Later, the bar codes can be used to rapidly weed out paper copies not required to be retained for tax or asset inventory purposes.
- If retailers put barcodes on receipts that allowed downloading an electronic copy or encoded the entire receipt in a 2D barcode, consumers could easily import data into personal finance, property inventory, and grocery management software. Receipts scanned on a scanner could be automatically identified and associated with the appropriate entries in finance and property inventory software.
- Consumer tracking from the retailer perspective (for example, loyalty card programs that track consumers purchases at the point of sale by having them scan a QR code).

A number of enterprise applications using cell phones are appearing:

- Access control (for example, ticket validation at venues), inventory reporting (for example, tracking deliveries), asset tracking (for example, anti-counterfeiting).

Smartphones

- Smartphones can be used in Google's mobile Android operating system via both their own Google Goggles application. Nokia's Symbian operating system features a barcode scanner which can scan barcodes, while mbarcode is a barcode reader for the Maemo operating system. In the Apple iOS, a barcode reader is not automatically included, but there are more than fifty free or paid apps available with both scanning capabilities and hard-linking to URI. With BlackBerry devices, the App World application can natively scan barcodes. Windows Phone 8 is able to scan barcodes through the Bing search app.

Housing

A large multifunction barcode scanner being used to monitor the transportation of packages of radioactive pharmaceuticals.

Barcode readers can be distinguished based on housing design as follows:

Handheld scanner

with a handle and typically a trigger button for switching on the light source.

Pen scanner (or wand scanner)

a pen-shaped scanner that is swiped.

Stationary scanner

Wall- Or Table-Mounted Scanners That The Barcode Is Passed Under Or Beside. These Are Commonly Found at the checkout counters of supermarkets and other retailers.

Fixed-position scanner

an industrial barcode reader, used to identify products during manufacture or logistics. Often used on conveyor tracks to identify cartons or pallets which need to be routed to another process or shipping location. Another application joins holographic scanners with a checkweigher to read bar codes of any orientation or placement, and weighs the package. Systems like this are used in factory and farm automation for quality management and shipping.

PDA scanner (or Auto-ID PDA)

a PDA with a built-in barcode reader or attached barcode scanner.

Automatic reader

a back office equipment to read barcoded documents at high speed (50,000/hour).

Cordless scanner (or Wireless scanner)

a cordless barcode scanner is operated by a battery fit inside it and is not connected to the electricity mains and transfer data to the connected device like PC.

Methods of Connection

Early Serial Interfaces

Early barcode scanners, of all formats, almost universally used the then-common RS-232 serial interface. This was an electrically simple means of connection and the software to access it is also relatively simple, although needing to be written for specific computers and their serial ports.

Proprietary Interfaces

There are a few other less common interfaces. These were used in large EPOS systems with dedicated hardware, rather than attaching to existing commodity computers. In some of these interfaces, the scanning device returned a "raw" signal proportional to the intensities seen while scanning the barcode. This was then decoded by the host device. In some cases the scanning device would convert the symbology of the barcode to one that could be recognized by the host device, such as Code 39.

Keyboard Wedge (e.g. PS/2)

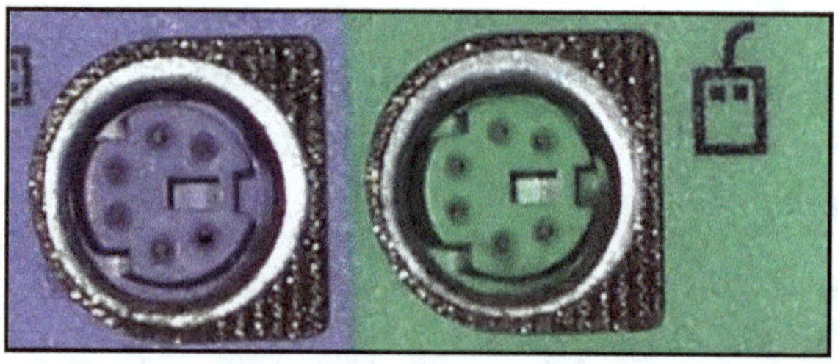

PS/2 keyboard and mouse ports.

With the popularity of the PC and its standard keyboard interface, it became ever easier to connect physical hardware to a PC and so there was commercial demand similarly to reduce the complexity of the associated software. "Keyboard wedge" (PS/2) hardware plugged between the PC and its normal keyboard, with characters from the barcode scanner appearing exactly as if they had been typed at the keyboard. This made the addition of simple barcode reading abilities to existing programs very easy, without any need to change them, although it did require some care by the user and could be restrictive in the content of the barcodes that could be handled.

USB

Later barcode readers began to use USB connectors rather than the keyboard port, as this became a more convenient hardware option. To retain the easy integration with existing programs, a device driver called a "software wedge" could be used, to emulate the keyboard-impersonating behavior of the old "keyboard wedge" hardware.

In many cases, a choice of USB interface types (HID, CDC) are provided. Some have PoweredUSB.

Wireless Networking

Some modern handheld barcode readers can be operated in wireless networks according to IEEE 802.11g (WLAN) or IEEE 802.15.1 (Bluetooth). Some barcode readers also support radio frequencies viz. 433 MHz or 910 MHz. Readers without external power sources require their batteries be recharged occasionally, which may make them unsuitable for some uses.

Resolution

The scanner resolution is measured by the size of the dot of light emitted by the reader. If this dot of light is wider than any bar or space in the bar code, then it will overlap two elements (two spaces or two bars) and it may produce wrong output. On the other hand, if a too small dot of light is used, then it can misinterpret any spot on the bar code making the final output wrong.

The most commonly used dimension is 13 thou (0.013 in or 0.33 mm), although some scanners can read codes with dimensions as small as 3 thou (0.003 in or 0.075 mm). Most manufacturers advertise bar code resolution in *mil, which is interchangeable with thou. Smaller bar codes must be printed at high resolution to be read accurately.*

Laser Surgery

Laser surgery is surgery using a laser (instead of a scalpel) to cut tissue. Examples include the use of a laser scalpel in otherwise conventional surgery, and soft-tissue laser surgery, in which the laser beam vaporizes soft tissue with high water content. Laser resurfacing is a technique in which covalent bonds of a material are dissolved by a laser, a technique invented by aesthetic plastic surgeon Thomas L Roberts, III using CO2 lasers in the 1990s. The CO2 (carbon dioxide) laser remains the gold standard for the soft tissue surgery because of the ease of simultaneous photo-thermal ablation and coagulation (and small blood capillary hemostasis).

Laser surgery is commonly used on the eye. Techniques used include LASIK, which is used to correct near and far-sightedness in vision, and photorefractive keratectomy, a procedure which permanently reshapes the cornea using an excimer laser to remove a small amount of tissue. Types of surgical lasers include carbon dioxide, argon, Nd:YAG laser, and Potassium titanyl phosphate.

Effects

1. Photochemical effect: clinically referred to as photodynamic therapy. Photosensitizer (photophrin II) is administered which is taken up by the tumor tissue and later irradiated by laser light resulting in highly toxic substances with resultant necrosis of the tumor. Photodynamic therapy is used in palliation of oesophagial and bronchial carcinoma and ablation of mucosal cancers of Gastrointestinal tract and urinary bladder.
2. Photoablative effect: Used in eye surgeries like band keratoplast, and endartectomy of peripheral blood vessels.
3. Photothermal effect: this property is used for endoscopic control of bleeding e.g. Bleeding peptic ulcers, oesophagial varices.
4. Photomechanical effect: used in intraluminal lithotripsy.

Applications

Dermatology and Plastic Surgery

A range of lasers such as erbium, dye, and CO_2 are used to treat various skin conditions including scars, vascular and pigmented lesions, and for photorejuvenation.

Eye Surgery

Various types of laser surgery are used to treat refractive error:

- ReLEx SMILE, use a femtosecond laser to create a refractive lenticule within the stroma which is then removed through a small incision.
- LASIK, in which a knife is used to cut a flap in the cornea, and a laser is used to reshape the layers underneath, to treat refractive error.
- IntraLASIK, a variant in which the flap is also cut with a laser.
- Photorefractive keratectomy (PRK, LASEK), in which the cornea is reshaped without first cutting a flap.
- Laser thermal keratoplasty, in which a ring of concentric burns is made in the cornea, which cause its surface to steepen, allowing better near vision.

Lasers are also used to treat non-refractive conditions, such as:

- Phototherapeutic keratectomy (PTK), in which opacities and surface irregularities are removed from the cornea.

- Laser coagulation, in which a laser is used to cauterize blood vessels in the eye, to treat various conditions.
- Lasers can be used to repair tears in the retina.

Endovascular Surgery

Laser endarterectomy is a technique in which an entire atheromatous plaque in the artery is excised. Laser recanalization of blocked arteries. other applications include laser assisted angioplasties and laser assisted vascular anastamosis.

Foot and Ankle Surgery

Lasers are used to treat several disorders in foot and ankle surgery. They are used to remove benign and malignant tumors, treat bunions, debride ulcers and burns, excise epidermal nevi, blue rubber bleb nevi, and keloids, and the removal of hypertrophic scars and tattoos.

A carbon dioxide laser (CO_2) is used in surgery to treat onychocryptosis (ingrown nails), onychauxis (club nails), onychogryposis (rams horn nail), and onychomycosis (fungus nail).

Gastro-Intestinal Tract

- Peritoneum-Laser is used for adhesiolysis.
- Peptic ulcer disease and oesophageal varices - Laser photoablation is done.
- Coagulation of vascular malformations of stomach, duodenum and colon.
- Lasers can be effectively used to treat early gastric cancers provided they are less than 4 cm and without lymph node involvement. Lasers are also used in treating oral submucous fibrosis.
- Palliative laser therapy is given in advanced oesophageal cancers with obstruction of lumen. Recanalisation of the lumen is done which allows the patient to resume soft diet and maintain hydration.
- Ablative laser therapy is used in advanced colorectal cancers to relieve obstruction and to control bleeding.
- Laser surgery used in hemorrhoidectomy, and is a relatively popular and non-invasive method of hemorrhoid removal.
- Laser-assisted liver resections have been done using carbon dioxide and Nd:YAG lasers.

- Ablation of liver tumors can be achieved by selective photovaporization of the tumor.
- Endoscopic laser lithotripsy is a safer modality compared to electrohydraulic lithotripsy.

Oral and Dental Surgery

The CO2 laser is used in oral and dental surgery for virtually all soft-tissue procedures, such as gingivecomies, vestibuloplasties, frenectomies and operculectomies. The CO2 10,600 nm wavelength is safe around implants as it is reflected by titanium, and thus has been gaining popularity in the field of periodontology. The laser may also be effective in treating peri-implantitis.

Other Surgery

The CO2 laser is also used in gynecology, genitourinary, general and thoracic surgery, otorhinolaryngology, orthopedic, and neurosurgery.

Fiber-Optic Communication

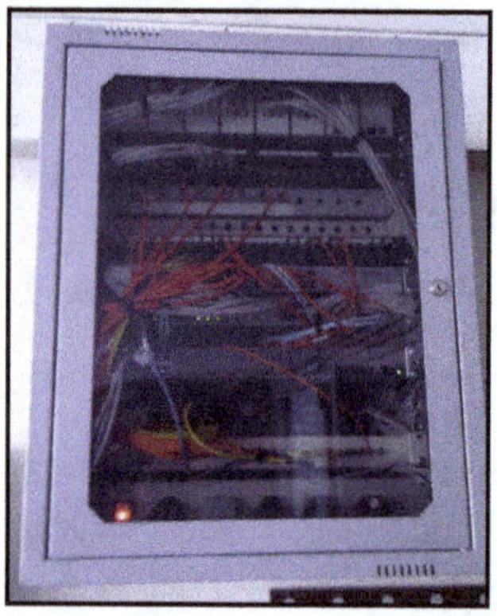

An optical fiber junction box. The yellow cables are single mode fibers; the orange and blue cables are multi-mode fibers: 62.5/125 µm OM1 and 50/125 µm OM3 fibers, respectively.

Stealth installing a 432-count dark fibre cable underneath the streets of Midtown Manhattan, New York City.

Fiber-optic communication is a method of transmitting information from one place to another by sending pulses of light through an optical fiber. The light forms an electromagnetic carrier wave that is modulated to carry information. First developed in the 1970s, fiber-optics have revolutionized the telecommunications industry and have played a major role in the advent of the Information Age. Because of its advantages over electrical transmission, optical fibers have largely replaced copper wire communications in core networks in the developed world. Optical fiber is used by many telecommunications companies to transmit telephone signals, Internet communication, and cable television signals. Researchers at Bell Labs have reached internet speeds of over 100 petabit×kilometer per second using fiber-optic communication.

The process of communicating using fiber-optics involves the following basic steps:

- creating the optical signal involving the use of a transmitter, usually from an electrical signal.
- relaying the signal along the fiber, ensuring that the signal does not become too distorted or weak.
- receiving the optical signal.
- converting it into an electrical signal.

Applications

Optical fiber is used by many telecommunications companies to transmit telephone signals, Internet communication, and cable television signals. Due to much lower attenuation and interference, optical fiber has large advantages over existing copper wire in long-distance and high-demand applications. However, infrastructure development within cities was relatively difficult and time-consuming, and fiber-optic systems were complex and expensive to install and operate. Due to these difficulties, fiber-optic communication systems have primarily been installed in long-distance applications, where they can be used to their full transmission capacity, offsetting the increased cost. Since 2000, the prices for fiber-optic communications have dropped considerably.

The price for rolling out fiber to the home has currently become more cost-effective than that of rolling out a copper based network. Prices have dropped to $850 per subscriber in the US and lower in countries like The Netherlands, where digging costs are low and housing density is high.

Since 1990, when optical-amplification systems became commercially available, the telecommunications industry has laid a vast network of intercity and transoceanic fiber communication lines. By 2002, an intercontinental network of 250,000 km of submarine communications cable with a capacity of 2.56 Tb/s was completed, and although specific network capacities are privileged information, telecommunications investment reports indicate that network capacity has increased dramatically since 2004.

History

In 1880 Alexander Graham Bell and his assistant Charles Sumner Tainter created a very early precursor to fiber-optic communications, the Photophone, at Bell's newly established Volta Laboratory in Washington, D.C. Bell considered it his most important invention. The device allowed for the transmission of sound on a beam of light. On June 3, 1880, Bell conducted the world's first wireless telephone transmission between two buildings, some 213 meters apart. Due to its use of an atmospheric transmission medium, the Photophone would not prove practical until advances in laser and optical fiber technologies permitted the secure transport of light. The Photophone's first practical use came in military communication systems many decades later.

In 1954 Harold Hopkins and Narinder Singh Kapany showed that rolled fiber glass allowed light to be transmitted. Initially it was considered that the light can traverse in only straight medium.

In 1966 Charles K. Kao and George Hockham proposed optical fibers at STC Laboratories (STL) at Harlow, England, when they showed that the losses of 1,000 dB/km in existing glass (compared to 5-10 dB/km in coaxial cable) was due to contaminants, which could potentially be removed.

Optical fiber was successfully developed in 1970 by Corning Glass Works, with attenuation low enough for communication purposes (about 20dB/km), and at the same time GaAs semiconductor lasers were developed that were compact and therefore suitable for transmitting light through fiber optic cables for long distances.

After a period of research starting from 1975, the first commercial fiber-optic communications system was developed, which operated at a wavelength around 0.8 μm and used GaAs semiconductor lasers. This first-generation system operated at a bit rate of 45 Mbps with repeater spacing of up to 10 km. Soon on 22 April 1977, General Telephone and Electronics sent the first live telephone traffic through fiber optics at a 6 Mbit/s throughput in Long Beach, California.

In October 1973, Corning Glass signed a development contract with CSELT and Pirelli aimed to test fiber optics in an urban environment: in September 1977, the second cable in this test series, named COS-2, was experimentally deployed in two lines (9 km) in Turin, for the first time in a big city, at a speed of 140 Mbit/s.

The second generation of fiber-optic communication was developed for commercial use in the early 1980s, operated at 1.3 μm, and used InGaAsP semiconductor lasers. These early systems were initially limited by multi mode fiber dispersion, and in 1981 the single-mode fiber was revealed to greatly improve system performance, however practical connectors capable of working with single mode fiber proved difficult to develop. In 1984, they had already developed a fiber optic cable that would help further their progress toward making fiber optic cables that would circle the globe. Canadian service provider SaskTel had completed construction of what was then the world's longest commercial fiberoptic network, which covered 3,268 km and linked 52 communities. By 1987, these systems were operating at bit rates of up to 1.7 Gb/s with repeater spacing up to 50 km.

The first transatlantic telephone cable to use optical fiber was TAT-8, based on Desurvire optimized laser amplification technology. It went into operation in 1988.

Third-generation fiber-optic systems operated at 1.55 μm and had losses of about 0.2 dB/km. This development was spurred by the discovery of Indium gallium arsenide and the development of the Indium Gallium Arsenide photodiode by Pearsall. Engineers overcame earlier difficulties with pulse-spreading at that wavelength using conventional InGaAsP semiconductor lasers. Scientists overcame this difficulty by using dispersion-shifted fibers designed to have minimal dispersion at 1.55 μm or by limiting the laser spectrum to a single longitudinal mode. These developments eventually allowed third-generation systems to operate commercially at 2.5 Gbit/s with repeater spacing in excess of 100 km.

The fourth generation of fiber-optic communication systems used optical amplification to reduce the need for repeaters and wavelength-division multiplexing to increase data capacity. These two improvements caused a revolution that resulted in the doubling of system capacity every six months starting in 1992 until a bit rate of 10 Tb/s was reached by 2001. In 2006 a bit-rate of 14 Tbit/s was reached over a single 160 km line using optical amplifiers.

The focus of development for the fifth generation of fiber-optic communications is on extending the wavelength range over which a WDM system can operate. The conventional wavelength window, known as the C band, covers the wavelength range 1.53-1.57 μm, and *dry fiber has a low-loss window promising an extension of that range to 1.30-1.65 μm. Other developments include the concept of* "optical solitons, " pulses that preserve their shape by counteracting the effects of dispersion with the nonlinear effects of the fiber by using pulses of a specific shape.

In the late 1990s through 2000, industry promoters, and research companies such as KMI, and RHK predicted massive increases in demand for communications bandwidth due to increased use of the Internet, and commercialization of various bandwidth-intensive consumer services, such as video on demand. Internet protocol data traffic was increasing exponentially, at a faster rate than integrated circuit complexity had increased under Moore's Law. From the bust of the dot-com bubble through 2006, however, the main trend in the industry has been consolidation of firms and offshoring of manufacturing to reduce costs. Companies such as Verizon and AT&T have taken advantage of fiber-optic communications to deliver a variety of high-throughput data and broadband services to consumers' homes.

Technology

Modern fiber-optic communication systems generally include an optical transmitter to convert an electrical signal into an optical signal to send into the optical fiber, a cable containing bundles of multiple optical fibers that is routed through underground conduits and buildings, multiple kinds of amplifiers, and an optical receiver to recover the signal as an electrical signal. The information transmitted is typically digital information generated by computers, telephone systems, and cable television companies.

Transmitters

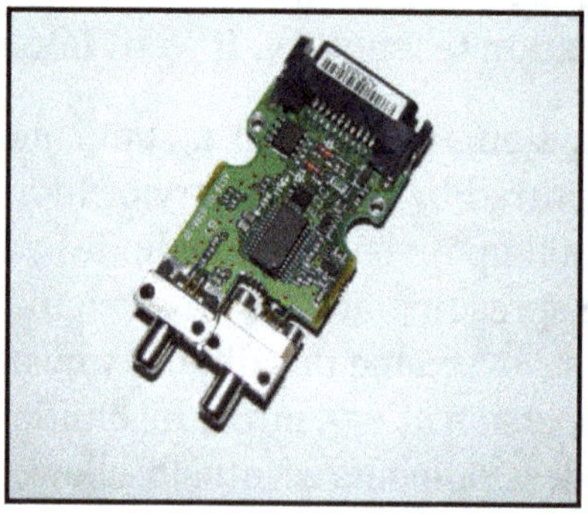

A GBIC module (shown here with its cover removed), is an optical and electrical transceiver. The electrical connector is at top right, and the optical connectors are at bottom left.

The most commonly used optical transmitters are semiconductor devices such as light-emitting diodes (LEDs) and laser diodes. The difference between LEDs and laser diodes is that LEDs produce incoherent light, while laser diodes produce coherent light. For use in optical communications, semiconductor optical transmitters must be designed to be compact, efficient, and reliable, while operating in an optimal wavelength range, and directly modulated at high frequencies.

In its simplest form, a LED is a forward-biased p-n junction, emitting light through spontaneous emission, a phenomenon referred to as electroluminescence. The emitted light is incoherent with a relatively wide spectral width of 30-60 nm. LED light transmission is also inefficient, with only about 1% of input power, or about 100 microwatts, eventually converted into launched power which has been coupled into the optical fiber. However, due to their relatively simple design, LEDs are very useful for low-cost applications.

Communications LEDs are most commonly made from Indium gallium arsenide phosphide (InGaAsP) or gallium arsenide (GaAs). Because InGaAsP LEDs operate at a longer wavelength than GaAs LEDs (1.3 micrometers vs. 0.81-0.87 micrometers), their output spectrum, while equivalent in energy is wider in wavelength terms by a factor of about 1.7. The large spectrum width of LEDs is subject to higher fiber dispersion, considerably limiting their bit rate-distance product (a common measure of usefulness). LEDs are suitable primarily for local-area-network applications with bit rates of 10-100 Mbit/s and transmission distances of a few kilometers. LEDs have also been developed that use several quantum wells to emit light at different wavelengths over a broad spectrum, and are currently in use for local-area WDM (Wavelength-Division Multiplexing) networks.

Today, LEDs have been largely superseded by VCSEL (Vertical Cavity Surface Emitting Laser) devices, which offer improved speed, power and spectral properties, at a similar cost. Common VCSEL devices couple well to multi mode fiber.

A semiconductor laser emits light through stimulated emission rather than spontaneous emission, which results in high output power (~100 mW) as well as other benefits related to the nature of coherent light. The output of a laser is relatively directional, allowing high coupling efficiency (~50 %) into single-mode fiber. The narrow spectral width also allows for high bit rates since it reduces the effect of chromatic dispersion. Furthermore, semiconductor lasers can be modulated directly at high frequencies because of short recombination time.

Commonly used classes of semiconductor laser transmitters used in fiber optics include VCSEL (Vertical-Cavity Surface-Emitting Laser), Fabry–Pérot and DFB (Distributed Feed Back).

Laser diodes are often directly modulated, that is the light output is controlled by a current applied directly to the device. For very high data rates or very long distance *links, a laser source may be operated* continuous wave, and the light modulated by an external device such as an electro-absorption modulator or Mach–Zehnder interferometer. External modulation increases the achievable link distance by eliminating laser chirp, which broadens the linewidth of directly modulated lasers, increasing the chromatic dispersion in the fiber.

A transceiver is a device combining a transmitter and a receiver in a single housing.

Receivers

The main component of an optical receiver is a photodetector, which converts light into electricity using the photoelectric effect. The primary photodetectors for telecommunications are made from Indium gallium arsenide The photodetector is typically a semiconductor-based photodiode. Several types of photodiodes include p-n photodiodes, p-i-n photodiodes, and avalanche photodiodes. Metal-semiconductor-metal (MSM) photodetectors are also used due to their suitability for circuit integration in regenerators and wavelength-division multiplexers.

Optical-electrical converters are typically coupled with a transimpedance amplifier and a limiting amplifier to produce a digital signal in the electrical domain from the incoming optical signal, which may be attenuated and distorted while passing through the channel. Further signal processing such as clock recovery from data (CDR) performed by a phase-locked loop may also be applied before the data is passed on.

Fiber Cable Types

An optical fiber cable consists of a core, cladding, and a buffer (a protective outer coating), in which the cladding guides the light along the core by using the method of total internal reflection. The core and the cladding (which has a lower-refractive-index) are usually made of high-quality silica glass, although they can both be made of plastic as well. Connecting two optical fibers is done by fusion splicing or mechanical splicing and requires special skills and interconnection technology due to the microscopic precision required to align the fiber cores.

A cable reel trailer with conduit that can carry optical fiber.

Multi-mode optical fiber in an underground service pit.

Two main types of optical fiber used in optic communications include multi-mode optical fibers and single-mode optical fibers. A multi-mode optical fiber has a larger core (≥ 50 micrometers), allowing less precise, cheaper transmitters and receivers to connect to it as well as cheaper connectors. However, a multi-mode fiber introduces multimode distortion, which often limits the bandwidth and length of the link. Furthermore, because of its higher dopant content, multi-mode fibers are usually expensive and exhibit higher attenuation. The core of a single-mode fiber is smaller (<10 micrometers) and requires more expensive components and interconnection methods, but allows much longer, higher-performance links.

In order to package fiber into a commercially viable product, it typically is protectively coated by using ultraviolet (UV), light-cured acrylate polymers, then terminated with optical fiber connectors, and finally assembled into a cable. After that, it can be laid in the ground and then run through the walls of a building and deployed aerially in a manner similar to copper cables. These fibers require less maintenance than common twisted pair wires, once they are deployed.

Specialized cables are used for long distance subsea data transmission, e.g. transatlantic communications cable. New (2011–2013) cables operated by commercial enterprises (Emerald Atlantis, Hibernia Atlantic) typically have four strands of fiber and cross the Atlantic (NYC-London) in 60-70ms. Cost of each such cable was about $300M in 2011. *source: The Chronicle Herald.*

Another common practice is to bundle many fiber optic strands within long-distance power transmission cable. This exploits power transmission rights of way effectively, ensures a power company can own and control the fiber required to monitor its own devices and lines, is effectively immune to tampering, and simplifies the deployment of smart grid technology.

Amplifier

The transmission distance of a fiber-optic communication system has traditionally been limited by fiber attenuation and by fiber distortion. By using opto-electronic repeaters, these problems have been eliminated. These repeaters convert the signal into an electrical signal, and then use a transmitter to send the signal again at a higher intensity than was received, thus counteracting the loss incurred in the previous segment. Because of the high complexity with modern wavelength-division multiplexed signals (including the fact that they had to be installed about once every 20 km), the cost of these repeaters is very high.

An alternative approach is to use an optical amplifier, which amplifies the optical signal directly without having to convert the signal into the electrical domain. It is made by doping a length of fiber with the rare-earth mineral erbium, and *pumping it with light from a* laser with a shorter wavelength than the communications signal (typically 980 nm). Amplifiers have largely replaced repeaters in new installations.

Wavelength-Division Multiplexing

Wavelength-division multiplexing (WDM) is the practice of multiplying the available capacity of optical fibers through use of parallel channels, each channel on a dedicated wavelength of light. This requires a wavelength division multiplexer in the transmitting equipment and a demultiplexer (essentially a spectrometer) in the receiving equipment. Arrayed waveguide gratings are commonly used for multiplexing and demultiplexing in WDM. Using WDM technology now commercially available, the bandwidth of a fiber can be divided into as many as 160 channels to support a combined bit rate in the range of 1.6 Tbit/s.

Parameters

Bandwidth–Distance Product

Because the effect of dispersion increases with the length of the fiber, a fiber transmission system is often characterized by its *bandwidth–distance product, usually expressed in units of* MHz·km. This value is a product of bandwidth and distance because there is a trade off between the bandwidth of the signal and the distance it can be carried. For example, a common multi-mode fiber with bandwidth–distance product of 500 MHz·km could carry a 500 MHz signal for 1 km or a 1000 MHz signal for 0.5 km.

Engineers are always looking at current limitations in order to improve fiber-optic communication, and several of these restrictions are currently being researched.

Record Speeds

Each fiber can carry many independent channels, each using a different wavelength of light (wavelength-division multiplexing). The net data rate (data rate without overhead bytes) per fiber is the per-channel data rate reduced by the FEC overhead, multiplied by the number of channels (usually up to eighty in commercial dense WDM systems as of 2008.

Year	Organization	Effective speed	WDM channels	Per channel speed	Distance
2009	Alcatel-Lucent	15 Tbit/s	155	100 Gbit/s	90 km
2010	NTT	69.1 Tbit/s	432	171 Gbit/s	240 km
2011	KIT	26 Tbit/s	1	26 Tbit/s	50 km
2011	NEC	101 Tbit/s	370	273 Gbit/s	165 km
2012	NEC, Corning	1.05 Petabit/s	12 core fiber		52.4 km

While the physical limitations of electrical cable prevent speeds in excess of 10 Gigabits per second, the physical limitations of fiber optics have not yet been reached.

In 2013, *New Scientist reported that a team at the* University of Southampton had achieved a throughput of 73.7 Tbit per second, with the signal traveling at 99.7% the speed of light through a hollow-core photonic crystal fiber.

Dispersion

For modern glass optical fiber, the maximum transmission distance is limited not by direct material absorption but by several types of dispersion, or spreading of optical pulses as they travel along the fiber. Dispersion in optical fibers is caused by a variety of factors. Intermodal dispersion, caused by the different axial speeds of different transverse modes, limits the performance of multi-mode fiber. Because single-mode fiber supports only one transverse mode, intermodal dispersion is eliminated.

In single-mode fiber performance is primarily limited by chromatic dispersion (also called group velocity dispersion), which occurs because the index of the glass varies slightly depending on the wavelength of the light, and light from real optical transmitters necessarily has nonzero spectral width (due to modulation). Polarization mode dispersion, another source of limitation, occurs because although the single-mode fiber can sustain only one transverse mode, it can carry this mode with two different polarizations, and slight imperfections or distortions in a fiber can alter the propagation velocities for the two polarizations. This phenomenon is called fiber birefringence and can be counteracted by polarization-maintaining optical fiber. Dispersion limits the bandwidth of the fiber because the spreading optical pulse limits the rate that pulses can follow one another on the fiber and still be distinguishable at the receiver.

Some dispersion, notably chromatic dispersion, can be removed by a 'dispersion compensator'. This works by using a specially prepared length of fiber that has the opposite dispersion to that induced by the transmission fiber, and this sharpens the pulse so that it can be correctly decoded by the electronics.

Attenuation

Fiber attenuation, which necessitates the use of amplification systems, is caused by a combination of material absorption, Rayleigh scattering, Mie scattering, and connection losses. Although material absorption for pure silica is only around 0.03 dB/km (modern fiber has attenuation around 0.3 dB/km), impurities in the original optical fibers caused attenuation of about 1000 dB/km. Other forms of attenuation are caused by physical stresses to the fiber, microscopic fluctuations in density, and imperfect splicing techniques.

Transmission Windows

Each effect that contributes to attenuation and dispersion depends on the optical wavelength. There are wavelength bands (or windows) where these effects are weakest, and these are the most favorable for transmission. These windows have been standardized, and the currently defined bands are the following:

Band	Description	Wavelength Range
O band	original	1260 to 1360 nm
E band	extended	1360 to 1460 nm
S band	short wavelengths	1460 to 1530 nm
C band	conventional ("erbium window")	1530 to 1565 nm
L band	long wavelengths	1565 to 1625 nm
U band	ultralong wavelengths	1625 to 1675 nm

Note that this table shows that current technology has managed to bridge the second and third windows that were originally disjoint.

Historically, there was a window used below the O band, called the first window, at 800-900 nm; however, losses are high in this region so this window is used primarily for short-distance communications. The current lower windows (O and E) around 1300 nm have much lower losses. This region has zero dispersion. The middle windows (S and C) around 1500 nm are the most widely used. This region has the lowest attenuation losses and achieves the longest range. It does have some dispersion, so dispersion compensator devices are used to remove this.

Regeneration

When a communications link must span a larger distance than existing fiber-optic technology is capable of, the signal must be *regenerated at intermediate points in the link by* optical communications repeaters. Repeaters add substantial cost to a communication system, and so system designers attempt to minimize their use.

Recent advances in fiber and optical communications technology have reduced signal degradation so far that *regeneration of the optical signal is only needed over distances of hundreds of kilometers. This has greatly reduced the cost of optical networking, particularly over undersea spans where the cost and reliability of repeaters is one of the key factors determining the performance of the whole cable system. The main advances contributing to these performance improvements are dispersion management, which seeks to balance the effects of dispersion against non-linearity; and* solitons, which use nonlinear effects in the fiber to enable dispersion-free propagation over long distances.

Last Mile

Although fiber-optic systems excel in high-bandwidth applications, optical fiber has been slow to achieve its goal of fiber to the premises or to solve the last mile problem. However, as bandwidth demand increases, more and more progress towards this goal can be observed. In Japan, for instance EPON has largely replaced DSL as a broadband Internet source. South Korea's KT also provides a service called FTTH (Fiber To The Home), which provides fiber-optic connections to the subscriber's home. The largest FTTH deployments are in Japan, South Korea, and China. Singapore started implementation of their all-fiber Next Generation Nationwide Broadband Network (Next Gen NBN), which is slated for completion in 2012 and is being installed by OpenNet. Since they began rolling out services in September 2010, Network coverage in Singapore has reached 85% nationwide.

In the US, Verizon Communications provides a FTTH service called FiOS to select high-ARPU (Average Revenue Per User) markets within its existing territory. The other major surviving ILEC (or Incumbent Local Exchange Carrier), AT&T, uses a FTTN (Fiber To The Node) service called U-verse with twisted-pair to the home. Their MSO competitors employ FTTN with coax using HFC. All of the major access networks use fiber for the bulk of the distance from the service provider's network to the customer.

Also in the US, Wilson Utilities, located in Wilson, North Carolina, has implemented FTTH and has successfully achieved 1 gigabit fiber to the home. This was implemented in late 2013. Wilson Utilities first rolled out their FTTH (Fiber to the Home) in 2012 with speeds offerings of 20/40/60/100 megabits per second. Their service is referred to as GreenLight.

Some other small cities in the US, such as Morristown, TN, have had their local utility company, Morristown Utility Systems in this case, deploy FTTH, offering symmetric gigabit speeds to each subscriber (though most are 50/50 or 100/100 mbit). It's called MUS Fibernet. AT&T and others have aggressively sought legislation at the state level to prevent further competition from municipalities, despite their low investment in rural areas.

The globally dominant access network technology is EPON (Ethernet Passive Optical Network). In Europe, and among telcos in the United States, BPON (ATM-based Broadband PON) and GPON (Gigabit PON) had roots in the FSAN (Full Service Access Network) and ITU-T standards organizations under their control.

Comparison with Electrical Transmission

The choice between optical fiber and electrical (or copper) transmission for a particular system is made based on a number of trade-offs. Optical fiber is generally chosen for systems requiring higher bandwidth or spanning longer distances than electrical cabling can accommodate.

A mobile fiber optic splice lab used to access and splice underground cables.

The main benefits of fiber are its exceptionally low loss (allowing long distances between amplifiers/repeaters), its absence of ground currents and other parasite signal and power issues common to long parallel electric conductor runs (due to its reliance on light rather than electricity for transmission, and the dielectric nature of fiber optic), and its inherently high data-carrying capacity. Thousands of electrical links would be required to replace a single high bandwidth fiber cable. Another benefit of fibers is that even when run alongside each other for long distances, fiber cables experience effectively no crosstalk, in contrast to some types of electrical transmission lines. Fiber can be installed in areas with high electromagnetic interference (EMI), such as alongside utility lines, power lines, and railroad tracks. Nonmetallic all-dielectric cables are also ideal for areas of high lightning-strike incidence.

For comparison, while single-line, voice-grade copper systems longer than a couple of kilometers require in-line signal repeaters for satisfactory performance; it is not unusual for optical systems to go over 100 kilometers (62 mi), with no active or passive processing. Single-mode fiber cables are commonly available in 12 km lengths, minimizing the number of splices required over a long cable run. Multi-mode fiber is available in lengths up to 4 km, although industrial standards only mandate 2 km unbroken runs.

An underground fiber optic splice enclosure opened up.

In short distance and relatively low bandwidth applications, electrical transmission is often preferred because of its:

- Lower material cost, where large quantities are not required.
- Lower cost of transmitters and receivers.
- Capability to carry electrical power as well as signals (in appropriately designed cables).
- Ease of operating transducers in linear mode.
- Crosstalk from nearby cables and other parasitical unwanted signals increase profits from replacement and mitigation devices.

Optical fibers are more difficult and expensive to splice than electrical conductors. And at higher powers, optical fibers are susceptible to fiber fuse, resulting in catastrophic destruction of the fiber core and damage to transmission components.

Because of these benefits of electrical transmission, optical communication is not common in short box-to-box, backplane, or chip-to-chip applications; however, optical systems on those scales have been demonstrated in the laboratory.

In certain situations fiber may be used even for short distance or low bandwidth applications, due to other important features:

- Immunity to electromagnetic interference, including nuclear electromagnetic pulses.
- High electrical resistance, making it safe to use near high-voltage equipment or between areas with different earth potentials.
- Lighter weight—important, for example, in aircraft.
- No sparks—important in flammable or explosive gas environments.
- Not electromagnetically radiating, and difficult to tap without disrupting the signal—important in high-security environments.
- Much smaller cable size—important where pathway is limited, such as networking an existing building, where smaller channels can be drilled and space can be saved in existing cable ducts and trays.
- Resistance to corrosion due to non-metallic transmission medium.

Optical fiber cables can be installed in buildings with the same equipment that is used to install copper and coaxial cables, with some modifications due to the small size and limited pull tension and bend radius of optical cables. Optical cables can typically be installed in duct systems in spans of 6000 meters or more depending on the duct's condition, layout of the duct system, and installation technique. Longer cables can be coiled at an intermediate point and pulled farther into the duct system as necessary.

Governing Standards

In order for various manufacturers to be able to develop components that function compatibly in fiber optic communication systems, a number of standards have been developed. The International Telecommunications Union publishes several standards related to the characteristics and performance of fibers themselves, including

- ITU-T G.651, "Characteristics of a 50/125 μm multimode graded index optical fibre cable".
- ITU-T G.652, "Characteristics of a single-mode optical fibre cable".

Other standards specify performance criteria for fiber, transmitters, and receivers to be used together in conforming systems. Some of these standards are:

- 100 Gigabit Ethernet.
- 10 Gigabit Ethernet.
- Fibre Channel.

- Gigabit Ethernet.
- HIPPI.
- Synchronous Digital Hierarchy.
- Synchronous Optical Networking.
- Optical Transport Network (OTN).

TOSLINK is the most common format for digital audio cable using plastic optical fiber to connect digital sources to digital receivers.

Laser Pointer

A laser pointer or laser pen is a small handheld device with a power source (usually a battery) and a laser diode emitting a very narrow coherent low-powered laser beam of visible light, intended to be used to highlight something of interest by illuminating it with a small bright spot of colored light. Power is restricted in most jurisdictions not to exceed 5 mW.

The small width of the beam and low power of typical laser pointers make the beam itself invisible in a reasonably clean atmosphere, only showing a point of light when striking an opaque surface. Some higher-powered laser pointers project a visible beam via scattering from dust particles or water droplets along the beam path. Higher-power and higher-frequency green or blue lasers may produce a beam visible even in clean air because of Rayleigh scattering from air molecules, especially when viewed in moderately-to-dimly lit conditions. The intensity of such scattering increases when these beams are viewed from angles near the beam axis. Such pointers, particularly in the green-light output range, are used as astronomical-object pointers for teaching purposes.

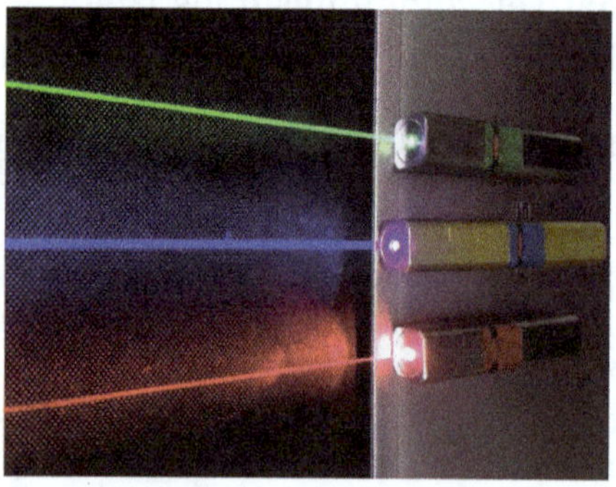

Red (635 nm), green (520 nm) and blue (445 nm) laser pointers.

The low-cost availability of infrared (IR) diode laser modules of up to 1000 mW (1 watt) output has created a generation of IR-pumped, frequency doubled, green, blue, and violet diode-pumped solid-state laser pointers with visible power up to 300 mW. Because the invisible IR component in the beams of these visible lasers is difficult to filter out, and also because filtering it contributes extra heat which is difficult to dissipate in a small pocket "laser pointer" package, it is often left as a beam component in cheaper high-power pointers. This invisible IR component causes a degree of extra potential hazard in these devices when pointed at nearby objects and people.

Laser pointers make a potent signaling tool, even in daylight, and are able to produce a bright signal for potential search and rescue vehicles using an inexpensive, small and lightweight device of the type that could be routinely carried in an emergency kit.

If aimed at a person's eyes, laser pointers can cause temporary disturbances to vision. There is some evidence of rare minor permanent harm, but low-powered laser pointers are not seriously hazardous to health. They may be a major annoyance in some circumstances. A dot of light from a red laser pointer may be thought to be due to a laser gunsight. When pointed at aircraft at night, laser pointers may dazzle and distract pilots, and increasingly strict laws have been passed to ban this.

Colors and Wavelengths

Early laser pointers were helium–neon (HeNe) gas lasers and generated laser radiation at 633 nanometers (nm), usually designed to produce a laser beam with an output power under 1 milliwatt (mW). The least expensive laser pointers use a deep-red laser diode near the 650 nm wavelength. Slightly more expensive ones use a red-orange 635 nm diode, more easily visible because of the greater sensitivity of the human eye at 635 nm. Other colors are possible too, with the 532 nm green laser being the most common alternative. Yellow-orange laser pointers, at 593.5 nm, later became available. In September 2005 handheld blue laser pointers at 473 nm became available. In early 2010 "Blu-ray" (actually violet) laser pointers at 405 nm went on sale.

The apparent brightness of a spot from a laser beam depends on the optical power of the laser, the reflectivity of the surface, and the chromatic response of the human eye. For the same optical power, green laser light will seem brighter than other colors because the human eye is most sensitive at low light levels in the green region of the spectrum (wavelength 520–570 nm). Sensitivity decreases for longer (redder) and shorter (bluer) wavelengths.

The output power of a laser pointer is usually stated in milliwatts (mW). In the U.S. lasers are classified by the American National Standards Institute and Food and Drug Administration (FDA). Visible laser pointers (400–700 nm) operating at less than 1 mW power are Class 2 or II, and visible laser pointers operating with 1–5 mW power are Class 3A or IIIa. Class 3B or IIIb lasers generate between 5 and 500 mW; Class 4 or IV lasers generate more than 500 mW. The US FDA Code of Federal Regulations stipulates that "demonstration laser products" such as pointers must comply with applicable requirements for Class I, IIa, II, or IIIa devices.

Red and Red-Orange

These are the simplest pointers, as laser diodes are available in these wavelengths. The pointer is nothing more than a battery-powered laser diode. The first red laser pointers released in the early 1980s were large, unwieldy devices that sold for hundreds of dollars. Today, they are much smaller and generally cost very little. In the 21st century, diode-pumped solid-state (DPSS) red laser pointers emitting at 671 nm became available. Although this wavelength can be obtained directly with an inexpensive laser diode, higher beam quality and narrower spectral bandwidth are achieved through DPSS versions.

Yellow

Yellow laser pointers emitting at 593.5 nm became available in the last few years[*when?*]. Although they are based on the DPSS process, in this case two lasing lines of the ND:YVO4, 1064 nm and 1342 nm, are summed together with a nonlinear crystal. The complexity of this process makes these laser pointers inherently unstable and inefficient, with their outputs ranging from 1 mW to about 10 mW, greatly varying with temperature and usually mode-hopping if they get too hot or too cold. That is because such a complex process may require temperature stabilizers and active cooling, which can't be mounted into a small-sized host. Also, most smaller 593.5 nm pointers work in pulsed mode, so they can use smaller and less powerful pumping diodes. New 589 nm yellow laser pointers have been introduced using a more robust and secretive method[*clarify*] of harmonic generation from a DPSS laser system. This "sodium" wavelength, although only 4.5 nm away from the older 593.5 nm, appears more gold in colour compared to the more amber appearance of the 593.5 nm wavelength. Astronomical observatories use a specially tuned dye laser at 589.2 nm (yellow) to create a laser guide star for use with astronomical adaptive optics.

Green

Trails by a 15 mW green laser pointer in a time exposure of a living room at night.

Green laser pointers appeared on the market around 2000 and are the most common type of DPSS lasers (also called *diode-pumped solid-state frequency-doubled, DPSSFD). They are more complicated than standard red laser pointers, because* laser diodes are not commonly available in this wavelength range. The green light is generated in an indirect process, beginning with a high-power (typically 100–300 mW) infrared aluminium gallium arsenide (AlGaAs) laser diode operating at 808 nm. The 808 nm light pumps a crystal of neodymium-doped yttrium orthovanadate (Nd:YVO4 or neodymium-doped yttrium aluminium garnet (Nd:YAG), or, less commonly, neodymium-doped yttrium lithium fluoride (Nd:YLF)), which lases deeper in the infrared at 1064 nm. This lasing action is due to an electronic transition in the fluorescent neodymium ion, Nd(III), which is present in all of these crystals.

The Nd:YVO4 or other Nd-doped crystal is coated on the diode side with a dielectric mirror that reflects at 808 nm and transmits at 1064 nm. The crystal is mounted on a copper block, acting as a heat sink; its 1064 nm output is fed into a crystal of potassium titanyl phosphate (KTP), mounted on a heat sink in the laser cavity resonator. The orientation of the crystals must be matched, as they are both anisotropic and the Nd:YVO4 outputs polarized light. This unit acts as a frequency doubler and halves the wavelength to the desired 532 nm. The resonant cavity is terminated by a dielectric mirror that reflects at 1064 nm and transmits at 532 nm. An infrared filter behind the mirror removes IR radiation from the output beam (this may be omitted or inadequate in less-expensive "pointer-style" green lasers), and the assembly ends in a collimator lens.

Nd:YVO4 is replacing other Nd-doped materials such as Nd:YAG and Nd:YLF in such systems because of lower dependency on the exact parameters of the pump diode (therefore allowing for higher tolerances), wider absorption band, lower lasing threshold, higher slope efficiency, linear polarization of output light, and single-mode output. For frequency doubling of higher-power lasers, lithium triborate (LBO) is used instead of KTP. Newer lasers use a composite Nd:YVO4/KTP crystal instead of two discrete ones.

Some green lasers operate in pulse or quasi-continuous wave (QCW) mode to reduce cooling problems and prolong battery life.

An announcement in 2009 of a direct green laser (which does not require doubling) promises much higher efficiencies and could foster the development of new color video projectors.

In 2012, Nichia and OSRAM developed and manufactured merchant high-power green laser diodes (515/520 nm), which can emit green laser directly.

Because even a low-powered green laser is visible at night through Rayleigh scattering from air molecules, this type of pointer is used by astronomers to easily point out stars and constellations. Green laser pointers can come in a variety of different output powers. The 5 mW green laser pointers (class IIIa) are the safest to use, and anything more powerful is usually not necessary for pointing purposes, since the beam is still visible in dark lighting conditions.

The United States Coast Guard requires their air crews to return to base if a green laser is pointed at them, and have their eyes examined for eye damage. People have been given up to five years in jail for aiming a green laser at an aircraft.

Blue

Blue laser pointers in specific wavelengths such as 473 nm usually have the same basic construction as DPSS green lasers. In 2006 many factories began production of blue laser modules for mass-storage devices, and these were used in laser pointers too. These were DPSS-type frequency-doubled devices. They most commonly emit a beam at 473 nm, which is produced by frequency doubling of 946 nm laser radiation from a diode-pumped Nd:YAG or Nd:YVO4 crystal (Nd-doped crystals usually produce a principal wavelength of 1064 nm, but with the proper reflective coating mirrors can be also made to lase at other "higher harmonic" non-principal neodymium wavelengths). For high output power, BBO crystals are used as frequency doublers; for lower powers, KTP is used. The Japanese company Nichia controlled 80% of the blue-laser-diode market in 2006.

Some vendors are now selling collimated diode blue laser pointers with measured powers exceeding 1,500 mW. However, since the claimed power of "laser pointer" products also includes the IR power (in DPSS technology only) still present in the beam (for reasons discussed below), comparisons on the basis of strictly visual-blue component from DPSS-type lasers remain problematic, and the information is often not available. Because of the higher neodymium harmonic used, and the lower efficiency of frequency-doubling conversion, the fraction of IR power converted to 473 nm blue laser light in optimally configured DPSS modules is typically 10–13%, about half that typical for green lasers (20–30%).

Blue lasers can also be fabricated directly with InGaN semiconductors, which produce blue light without frequency doubling. 450 nm (447 nm plus/minus 5 nm) blue laser diodes are currently available on the open market. Some blue diodes are capable of very high power; such as Nichia's NDB7K75 diode, which can continuously output over 5 watts of energy if overdriven. The devices are brighter for the same power than 405 nm violet laser diodes, since the longer wavelength is closer to the peak sensitivity of the human eye. Mass production of laser diodes for commercial devices like laser projectors have driven down prices. Recent popularity of the high-power version of these 447 nm pointers, which also have improved optics for better collimation and lower divergence, rivals the hazards associated with the use of these portable devices by persons of questionable intention and cost has diminished to be competitive with DPSS green lasers wavelengths.

Violet

Lasers emitting a violet light beam at 405 nm may be constructed with GaN (gallium nitride) semiconductors. This is close to ultraviolet, bordering on the very extreme of human vision, and can cause bright blue fluorescence, and thus a blue rather than violet spot, on many white surfaces, including white clothing, white paper, and projection screens, due to the widespread use of optical brighteners in the manufacture of products intended to appear brilliantly white. On ordinary non-fluorescent materials, and also on fog or dust, the color appears as a shade of deep violet that cannot be reproduced on monitors and print. A GaN laser emits 405 nm directly without a frequency doubler, eliminating the possibility of accidental dangerous infrared emission. These laser diodes are mass-produced for the reading and writing of data in Blu-ray drives (although the light emitted by the diodes is not blue, but distinctly violet). As of September 2011[update], 405 nm blue-violet laser diode modules with an optical power of 250 mW, based on GaN violet laser diodes made for Blu-ray disc readers, had reached the market from Chinese sources for prices of about US$60 including delivery.

At the same time, a few higher-powered (120 mW) 404–405 nm "violet" laser pointers have become available that are not based on GaN, but use DPSS frequency-doubler technology from 1-watt 808 nm GaAlAs infrared diode lasers. As with infrared-driven green laser pointers above, such devices are able to pop balloons and light matches, but this is as a result of an unfiltered high-power infrared component in the beam.

Applications

Pointing

A 5 mW green laser pointer directed at a palm tree at night. Note that the beam itself is visible through Rayleigh scattering.

Laser pointers are often used in educational and business presentations and visual demonstrations as an eye-catching pointing device. Laser pointers enhance verbal guidance given to students during surgery. The suggested mechanism of explanation is that the technology enables a more precise guidance of location and identification of anatomic structures.

Red laser pointers can be used in almost any indoor or low-light situation where pointing out details by hand may be inconvenient, such as in construction work or interior decorating. Green laser pointers can be used for similar purposes as well as outdoors in daylight or for longer distances.

Laser pointers are used in a wide range of applications. Green laser pointers can also be used for amateur astronomy. Green laser is visible at night due to Rayleigh scattering and airborne dust, allowing someone to point out individual stars to others nearby. Also, these green laser pointers are commonly used by astronomers worldwide at star parties or for conducting lectures in astronomy. Astronomy laser pointers are also commonly mounted on telescopes in order to align the telescope to a specific star or location. Laser alignment is much easier than aligning through using the eyepiece.

Industrial and Research Use

Laser pointers are used in industry. For instance, construction companies may use high quality laser pointers to enhance the accuracy of showing specific distances, while working on large-scale projects. They proved to be useful in this type of business because of their accuracy, which made them significant time-savers. What is essentially a laser pointer may be built into an infrared thermometer to identify where it is pointing, or be part of a laser level or other apparatus.

They may also be helpful in scientific research in fields such as photonics, chemistry, physics, and medicine.

Laser pointers are used in robotics, for example, for laser guidance to direct the robot to a goal position by means of a laser beam, i.e. showing goal positions to the robot optically instead of communicating them numerically. This intuitive interface simplifies directing the robot while visual feedback improves the positioning accuracy and allows for implicit localization.

Leisure and Entertainment

Entertainment is one of the other applications that have been found for lasers. The most common use of lasers in entertainment can be seen in special effects used in laser shows. Clubs, parties and outdoor concerts all use high-power lasers, with safety precautions, as a spectacle. Laser shows are often extravagant, using lenses, mirrors and smoke.

Lasers have also become a popular plaything for pets and other animals, namely cats and dogs, whose natural prey drive is triggered by the moving laser and will chase it and/or try to catch it as much as possible, but obviously never succeed. As a result, laser pointers have become a popular form of entertainment for many pet owners.

However, laser *pointers have few applications beyond actual pointing in the wider entertainment industry, and many venues ban entry to those in possession of pointers as a potential hazard. Very occasionally laser gloves, which are sometimes mistaken for pointers, are seen being worn by professional dancers on stage at shows. Unlike pointers, these usually produce low-power highly divergent beams to en-*

sure eye safety. Laser pointers have been used as props by magicians during magic shows.

As an example of the potential dangers of laser pointers brought in by audience members, at the Tomorrow Land Festival in Belgium in 2009, laser pointers brought in by members of the audience of 200 mW or greater were found to be the cause of eye damage suffered by several other members of the audience according to reports about the incident filed on the ILDA (International Laser Display Association's) Web site. The report says that the incident was investigated by several independent authorities, including the Belgium police, and that those authorities concluded that pointers brought in by the audience were the cause of the injuries.

Laser pointers can be used in hiking or outdoor activities. Higher-powered laser pointers are bright enough to scare away large wild animals which makes them useful for hiking and camping. In these circumstances a laser pointer can also serve as a handy survival tool, as it can be used as a rescue signal in emergencies which is visible to aircraft and other parties, during both day and night conditions, at extreme distances. For example, during the night in August 2010 two men and a boy were rescued from marshland after their red laser pen was spotted by rescue teams.

Weapons Systems

Accurately aligned laser pointers are used as laser gunsights to aim a firearm.

Some militaries use lasers to mark targets at night for aircraft. This is done to ensure that "friendly" and "enemy" targets are not mistaken. A friendly target may wear an IR emitting device that is only visible to those utilizing night vision (such as pilots.) To pinpoint the exact location of an enemy combatant, they would simply illuminate the target with a laser beam detectable by the attacking aircraft. This can be one of the most accurate ways of marking targets.

Hazards

Laser pointers can cause eye injuries directly, and dangerous situations by distraction. As of 2011[update], in spite of the very large number of pointers, and many incidents of malicious or careless use, no very serious and permanent eye injuries have been reported from low-power pointers, nor have any aviation or other accidents actually been reported to be caused by them, although the risk is clearly real.

Incorrect Power Rating

National Institute of Standards and Technology tests conducted on laser pointers labeled as Class IIIa or 3R in 2013 showed that about half of them emitted power at twice the Class limit, making their correct designation Class IIIb – more hazardous than Class IIIa. The highest measured power output was 66.5 milliwatts; more than

10 times the limit. Green laser light is generated from an infrared laser beam, which should be confined within the laser housing; however, more than 75% of the devices tested were found to emit infrared light in excess of the limit.

Malicious Use

Laser pointers, with their very long range, are often maliciously shone at people to distract or annoy them. This is considered particularly hazardous in the case of aircraft pilots, who may be dazzled or distracted at critical times. According to an MSNBC report there were over 2,836 incidents logged in the US by the FAA in 2010. Illumination by handheld green lasers is particularly serious, as the wavelength (532 nm) is near peak sensitivity of the dark-adapted eye and may appear to be 35 times brighter than a red laser of identical power output.

Irresponsible use of laser pointers is often frowned upon by members of the laser projector community who fear that their misuse may result in legislation affecting lasers designed to be placed within projectors and used within the entertainment industry. Others involved in activities where dazzling or distraction are dangerous are also a concern.

Another distressing and potentially dangerous misuse of laser pointers is to use them when the dot may reasonably be mistaken for that of a laser gunsight. Armed police have drawn their weapons in such circumstances.

Eye Injury

The output of laser pointers available to the general public is limited (and varies by country) in order to prevent accidental damage to the retina of human eyes. The U.K. Health Protection Agency recommended that "laser pointers generally available to the public should be restricted to less than 1 milliwatt as no injuries [like the one reported below to have caused retinal damage] have been reported at this power". In the U.S., regulatory authorities allow lasers up to 5 mW.

Studies have found that even low-power laser beams of not more than 5 mW can cause permanent retinal damage if gazed at for several seconds; however, the eye's blink reflex makes this highly unlikely. Such laser pointers have reportedly caused afterimages, flash blindness and glare, but not permanent damage, and are generally safe when used as intended.

A high-powered green laser pointer bought over the Internet was reported in 2010 to have caused a decrease of visual acuity from 6/6 to 6/12 (20/20 to 20/40); after two months acuity recovered to 6/6, but some retinal damage remained. The US FDA issued a warning after two anecdotal reports it received of eye injury from laser pointers.

Laser pointers available for purchase online can be capable of significantly higher power output than the pointers typically available in stores. Dubbed "Burning Lasers", these are designed to burn through light plastics and paper, and can have very similar external appearances to their low-power counterparts.

Studies in the early twenty-first century found that the risk to the human eye from accidental exposure to light from commercially available class IIIa laser pointers having powers up to 5 mW seemed rather small; however, prolonged viewing, such as deliberate staring into the beam for 10 or more seconds, can cause damage.

The UK Health Protection Agency warns against the higher-power typically green laser pointers available over the Internet, with power output of up to a few hundred milliwatts, as "extremely dangerous and not suitable for sale to the public."

Infrared Hazards of Diode-Pumped Solid-State Laser Pointers

Lasers classified as *pointers are intended to have outputs less than 5 mW total power* (Class 3R). At such power levels, an IR filter for a DPSS laser may not be required as the infrared (IR) output is relatively low and the brightness of the visible wavelength of the laser will cause the eye to react (blink reflex). However, higher-powered (> 5 mW) DPSS-type laser pointers have recently become available, usually through sources that do not follow laser safety regulations for laser packaging and labeling. These higher-powered lasers are often packaged in the same pointer-style housings as regular laser pointers, and usually lack the IR filters found in professional high-powered DPSS lasers, because of costs and additional efforts needed to accommodate them.

Though the IR from a DPSS laser is less collimated, the typical neodymium-doped crystals in such lasers do produce a true IR laser beam. The eye will usually react to the higher-powered visible light; however in higher power DPSS lasers the IR laser output can be significant. What poses a special hazard for this unfiltered IR output is its presence in conjunction with laser safety goggles designed to only block the visible wavelengths of the laser. Red goggles, for example, will block most green light from entering the eyes, but will pass IR light. The reduced light behind the goggles may also cause the pupils to dilate, increasing the hazard to the invisible IR light. Dual-frequency so-called YAG laser eyewear is significantly more expensive than single frequency laser eyewear, and is often not supplied with unfiltered DPSS pointer style lasers, which output 1064 nm IR laser light as well. These potentially hazardous lasers produce little or no visible beam when shone through the eyewear supplied with them, yet their IR-laser output can still be easily seen when viewed with an IR-sensitive video camera.

In addition to the safety hazards of unfiltered IR from DPSS lasers, the IR component may be inclusive of total output figures in some laser pointers.

Though green (532 nm) lasers are most common, IR filtering problems may also exist in other DPSS lasers, such as DPSS red (671 nm), yellow (589 nm) and blue (473 nm) lasers. These DPSS laser wavelengths are usually more exotic, more expensive, and generally manufactured with higher quality components, including filters, unless they are put into laser pointer style pocket-pen packages. Most red (635 nm, 660 nm), violet (405 nm) and darker blue (445 nm) lasers are generally built using dedicated laser diodes at the output frequency, not as DPSS lasers. These diode-based visible lasers do not produce IR light.

Regulations and Misuse

Since laser pointers became readily available, they have been misused, leading to the development of laws and regulations specifically addressing use of such lasers. Their very long range makes it difficult to find the source of a laser spot. In some circumstances they make people fear they are being targeted by weapons, as they are indistinguishable from dot type laser reticles. The very bright, small spot makes it possible to dazzle and distract drivers and aircraft pilots, and they can be dangerous to sight if aimed at the eyes.

In 1998, an audience member shone a laser at Kiss drummer Peter Criss's eyes while the band was performing Beth. After performing the song, Criss nearly stormed off the stage, and lead singer Paul Stanley challenged whoever shone the laser to fight him on stage:

In every crowd, there's one or two people who don't belong here, Now I know you want to bring it to school tomorrow when you go to sixth grade, but leave it at home when you go to the show.

—Paul Stanley,

In January 2005 a New Jersey man named David Banach was arrested for pointing a green laser pointer at a small jet flying overhead.

In football a laser pointer is a prohibited item at stadiums during FIFA tournaments and matches, according to FIFA Stadium Safety and Security Regulations, it is also prohibited in matches and competitions organised by UEFA. In 2008 laser pointers were aimed at players' eyes in a number of sport matches worldwide. Olympique Lyonnais was fined by UEFA because of a laser pointer beam aimed by a Lyon fan at Cristiano Ronaldo. In a World Cup final qualifier match held in Riyadh, Saudi Arabia between the home team and the South Korean team, South Korean goalkeeper Lee Woon-Jae was hit in the eye with a green laser beam. At the 2014 World Cup during the final group stage match between Algeria and Russia a green laser beam was directed on the face of Russian goalkeeper Igor Akinfeev. After the match the Algerian Football Federation was fined CHF50,000 (approx. £33,000/€41,100/US$56,200) by FIFA for the use of lasers and other violations of the rules by Algerian fans at the stadium.

In 2009 police in the United Kingdom began tracking the sources of lasers being shone at helicopters at night, logging the source using GPS, using thermal imaging cameras to see the suspect, and even the warm pointer if discarded, and calling in police dog teams. As of 2010 the penalty could be five years' imprisonment.

Despite legislation limiting the output of laser pointers in some countries, higher-power devices are currently produced in other regions (especially China and Hong Kong), and are frequently imported by customers who purchase them directly via Internet mail order. The legality of such transactions is not always clear; typically, the lasers are sold as research or OEM devices (which are not subject to the same power restrictions), with a disclaimer that they are not to be used as pointers. DIY videos are also often posted on Internet video sharing sites like YouTube which explain how to make a high-power laser pointer using the diode from an optical disc burner. As the popularity of these devices increased, manufacturers began manufacturing similar high-powered pointers. Warnings have been published on the dangers of such high-powered lasers. Despite the disclaimers, such lasers are frequently sold in packaging resembling that for laser pointers. Lasers of this type may not include safety features sometimes found on laser modules sold for research purposes.

There have been many incidents regarding, in particular, aircraft, and the authorities in many countries take them extremely serious. Many people have been convicted and sentenced, sometimes to several years' imprisonment.

Australia

In April 2008, citing a series of coordinated attacks on passenger jets in Sydney, the Australian government announced that it would restrict the sale and importation of certain laser items. The government had yet to determine which classes of laser pointers to ban. After some debate, the government voted to ban importation of lasers that emit a beam stronger than 1 mW, effective from July 1, 2008. Those whose professions require the use of a laser can apply for an exemption. In Victoria and the Australian Capital Territory a laser pointer with an accessible emission limit greater than 50 mW is classified as a prohibited weapon and any sale of such items must be recorded. In Western Australia, regulatory changes have classified laser pointers as controlled weapons and demonstration of a lawful reason for possession is required. The WA state government has also banned as of 2000 the manufacture, sale and possession of laser pointers higher than class 2. In New South Wales and the Australian Capital Territory the product safety standard for laser pointers prescribes that they must be a Class 1 or a Class 2 laser product. In February 2009 South African cricketer Wayne Parnell had a laser pointer directed at his eyes when attempting to take a catch, which he dropped. He denied that it was a reason for dropping the ball, but despite this the MCG decided to keep an eye out for the laser pointers. The laser pointer ban only applies to hand-held battery-powered laser devices and not laser modules.

In November 2015 a 14-year-old Tasmanian boy damaged both his eyes after shining a laser pen "... in his eyes for a very brief period of time". He burned his retinas near the macular, the area where most of a persons central vision is located. As a result, the boy has almost immediately lost 75% of his vision, with little hope of recovery.

Canada

New regulations controlling the importation and sale of laser pointers (portable, battery-powered) have been established in Canada in 2011 and are governed by Health Canada using the Consumer Protection Act for the prohibition of sale of Class 3B (IEC) or higher power lasers to "consumers" as defined in the Consumer Protection Act . Canadian federal regulation follows FDA (US Food & Drug Administration) CDRH, and IEC (International Electrotechnical Commission) hazard classification methods where manufacturers comply with the Radiation Emitting Devices Act. As of July 2011 three people had been charged under the federal Aeronautics Act, which carries a maximum penalty of $100,000 and five years in prison, for attempting to dazzle a pilot with a laser. Other charges that could be laid include mischief and assault.

Netherlands

Before 1998 Class 3A lasers were allowed. In 1998 it became illegal to trade Class 2 laser pointers that are "gadgets" (e.g. ball pens, key chains, business gifts, devices that will end up in children's possession, etc.). It is still allowed to trade Class 2 (< 1 mW) laser pointers proper, but they have to meet requirements regarding warnings and instructions for safe use in the manual. Trading of Class 3 and higher laser pointers is not allowed.

Sweden

The use of pointers with output power > 1 mW is regulated in public areas and school yards. From 1 January 2014 it is necessary to have a special permit in order to own a laser pointer with a classification of 3R, 3B or 4, i.e. over 1 mW.

United Kingdom

UK and most of Europe are now harmonized on Class 2 (<1 mW) for General presentation use laser pointers or laser pens. Anything above 1 mW is illegal for sale in the UK (import is unrestricted). Health and Safety regulation insists on use of Class 2 anywhere the public can come in contact with indoor laser light, and the DTI have urged Trading Standards authorities to use their existing powers under the General Product Safety Regulations 2005 to remove lasers above class 2 from the general market.

Since 2010, it is an offence in the UK to shine a light at an aircraft in flight so as to dazzle the pilot, whether intentionally or not, with a maximum penalty of a level 4 fine (currently £2500). It is also an offence to negligently or recklessly endanger an aircraft, with a maximum penalty of five years imprisonment and/or an unlimited fine.

To assist with enforcement, police helicopters use GPS and thermal imaging camera, together with dog teams on the ground, to help locate the offender; the discarded warm laser pointer is often visible on the thermal camera, and its wavelength can be matched to that recorded by an event recorder in the helicopter.

In 2014, a Flintshire 22-year-old was convicted of recklessly endangering the lives of a police search helicopter crew through the use of such a device, and awarded a five-month suspended sentence.

United States

Laser pointers are Class II or Class IIIa devices, with output beam power less than 5 milliwatts (<5 mW). According to U.S. Food and Drug Administration (FDA) regulations, more powerful lasers may not be sold or promoted as laser pointers. Also, any laser with class higher than IIIa (more than 5 milliwatts) requires a key-switch interlock and other safety features. Shining a laser pointer of any class at an aircraft is illegal and punishable by a fine of up to $11,000.

All laser products offered in commerce in the US must be registered with the FDA, regardless of output power.

In Utah it is a class C misdemeanor to point a laser pointer at a law enforcement officer and is an infraction to point a laser pointer at a moving vehicle.

In Arizona it is a Class 1 misdemeanor if a person "aims a laser pointer at a police officer if the person intentionally or knowingly directs the beam of light from an operating laser pointer at another person and the person knows or reasonably should know that the other person is a police officer." (Arizona Revised Statutes §13-1213)

On April 30, 2010, Clint Jason Brenner, 36, of Prescott, AZ was found guilty of two counts of endangerment, each a class 6 felony, and it was also found that each was a dangerous offense, for shining a handheld laser pointer at an Arizona Department of Public Safety helicopter in December 2009. He was given a sentence of two years in prison for each count, to run concurrently.

On November 2, 2009, Dana Christian Welch of Southern California was sentenced to 2.5 years in a federal prison after being found guilty of shining a hand-held laser light into the eyes of two pilots landing Boeing jets at John Wayne Airport.

Michigan

Public act 257 of 2003 makes it a felony for a person to "manufacture, deliver, possess, transport, place, use, or release" a "harmful electronic or electromagnetic device" for "an unlawful purpose"; also made into a felony is the act of causing "an individual to falsely believe that the individual has been exposed to a... harmful electronic or electromagnetic device."

Public act 328 of 1931 makes it a felony for a person to "sell, offer for sale, or possess" a "portable device or weapon from which an electric current, impulse, wave, or beam may be directed" and is designed "to incapacitate temporarily, injure, or kill".

Maine

Public law 264, H.P. 868 - L.D. 1271 criminalizes the knowing, intentional, and/or reckless use of an electronic weapon on another person, defining an electronic weapon as a portable device or weapon emitting an electric current, impulse, beam, or wave with disabling effects on a human being.

Massachusetts

Chapter 170 of the Acts of 2004, Section 140 of the General Laws, section 131J states: "No person shall possess a portable device or weapon from which an electric current, impulse, wave or beam may be directed, which current, impulse, wave or beam is designed to incapacitate temporarily, injure or kill, except ... Whoever violates this section shall be punished by a fine of not less than $500 nor more than $1,000 or by imprisonment in the house of correction for not less than 6 months nor more than 2 1/2 years, or by both such fine and imprisonment."

Magneto-Optical Trap

A magneto-optical trap (abbreviated MOT) is an apparatus that uses laser cooling with magneto-optical trapping in order to produce samples of cold, trapped, neutral atoms at temperatures as low as several microkelvins, two or three times the recoil limit. By combining the small momentum of a single photon with a velocity and spatially dependent absorption cross section and a large number of absorption-spontaneous emission cycles, atoms with initial velocities of hundreds of metres per second can be slowed to tens of centimetres per second.

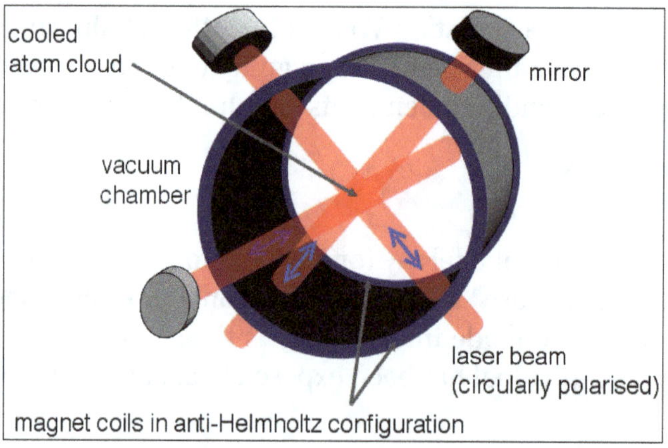

Experimental setup of the MOT.

Although charged particles can be trapped using a Penning trap or a Paul trap using a combination of electric and magnetic fields, those traps do not work for neutral atoms.

Doppler Cooling

Photons have a momentum given by $\hbar k$ (where $\hbar$ is the reduced Planck constant and k the photon wavenumber), which is conserved in all atom-photon interactions. Thus, when an atom absorbs a photon, it is given a momentum kick in the direction of the photon before absorption. By detuning a laser beam to a frequency less than the resonant frequency (also known as red detuning), laser light is only absorbed if the light is frequency up-shifted by the Doppler effect, which occurs whenever the atom is moving towards the laser source. This applies a friction force to the atom whenever it moves towards a laser source.

For cooling to occur along all directions, the atom must see this friction force along all three Cartesian axes; this is most easily achieved by illuminating the atom with 3 orthogonal laser beams, which are then reflected back along the same direction.

Magnetic Trapping

Magnetic trapping is created by adding a spatially varying magnetic quadrupole field to the red detuned optical field needed for laser cooling. This causes a Zeeman shift in the magnetic-sensitive mf levels, which increases with the radial distance from the centre of the trap. Because of this, as an atom moves away from the centre of the trap, the atomic resonance is shifted closer to the frequency of the laser light, and the atom becomes more likely to get a photon kick towards the centre of the trap.

The direction of the kick is given by the polarisation of the light, which is either left or right handed circular, giving different interactions with the different mf levels. The correct polarisations are used so that photons moving towards the centre of the trap will be on resonance with the correct shifted atomic energy level, always driving the atom towards the centre.

Atomic Structure Necessary for Magneto-Optical Trapping

As a thermal atom at room temperature has many thousands of times the momentum of a single photon, the cooling of an atom must involve many absorption-spontaneous emission cycles, with the atom losing up to ħk of momenta each cycle . Because of this, if an atom is to be laser cooled, it must possess a specific energy level structure known as a closed optical loop, where following an excitation-spontaneous emission event, the atom is always returned to its original state. 85Rubidium, for example, has a closed optical loop between the $5S_{1/2}$ $F=3$ state and the $5P_{3/2}$ $F=4$ state. Once in the excited state, the atom is forbidden from decaying to any of the $5P_{1/2}$ states, which would not conserve parity, and is also forbidden from decaying to the $5S_{1/2}$ $F=2$ state, which would require an angular momentum change of −2, which cannot be supplied by a single photon.

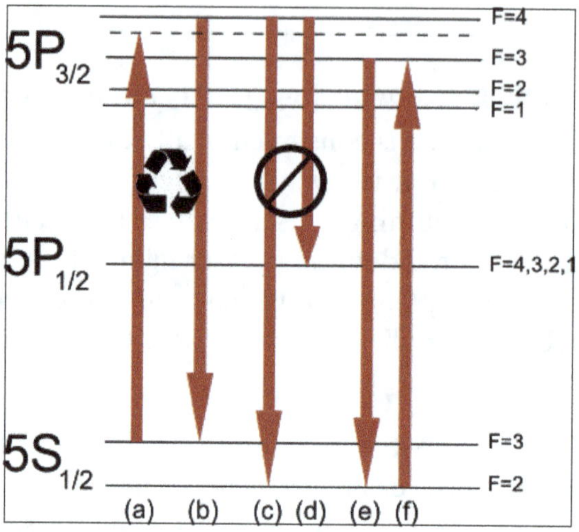

The lasers needed for the magneto-optical trapping of rubidium 85: (a) & (b) show the absorption (red detuned to the dotted line) and spontaneous emission cycle, (c) & (d) are forbidden transitions, (e) shows that if the cooling laser excites an atom to the $F=3$ state, it is allowed to decay to the "dark" lower hyperfine, *F=2 state, which would stop the cooling process, if it were not for the repumper laser (f).*

Many atoms that do not contain closed optical loops can still be laser cooled, however, by using repump lasers which re-excite the population back into the optical loop after it has decayed to a state outside of the cooling cycle. The magneto-optical trapping of rubidium 85, for example, involves cycling on the closed $5S_{1/2}\ F=3 \rightarrow 5P_{3/2}\ F=4$ transition. On excitation, however, the detuning necessary for cooling gives a small, but non-zero overlap with the $5P_{3/2}\ F=3$ state. If an atom is excited to this state, which occurs roughly every thousand cycles, the atom is then free to decay either the $F=3$, light coupled upper hyperfine state, or the $F=2$ "dark" lower hyperfine state. If it falls back to the dark state, the atom stops cycling between ground and excited state, and the cooling and trapping of this atom stops. A repump laser which is resonant with the $5S_{1/2}\ F=2 \rightarrow 5P_{3/2}\ F=3$ transition is used to recycle the population back into the optical loop so that cooling can continue.

Apparatus

Laser

All magneto-optical traps require at least one trapping laser plus any necessary repumper lasers. These lasers need stability, rather than high power, requiring no more than the saturation intensity, but a linewidth much less than the Doppler width, usually several megahertz. Because of their low cost, compact size and ease of use, laser diodes are used for many of the standard MOT species while the linewidth and stability of these lasers is controlled using servo systems, which stabilises the lasers to an atomic frequency reference by using, for example, saturated absorption spectroscopy and the Pound-Drever-Hall technique to generate a locking signal.

By employing a 2-dimensional diffraction grating it is possible to generate the configuration of laser beams required for a magneto-optical trap from a single laser beam and thus have a very compact magneto-optical trap.

Vacuum Chamber

The MOT cloud is loaded from a background of thermal vapour, or from an atomic beam, usually slowed down to the capture velocity using a Zeeman slower. However, the trapping potential in a magneto-optical trap is small in comparison to thermal energies of atoms and most collisions between trapped atoms and the background gas supply enough energy to the trapped atom to kick it out of the trap. If the background pressure is too high, atoms are kicked out of the trap faster than they can be loaded, and the trap does not form. This means that the MOT cloud only forms in a vacuum chamber with a background pressure of less than 10 micropascals (10−10 bar).

The Limits to The Magneto-Optical Trap

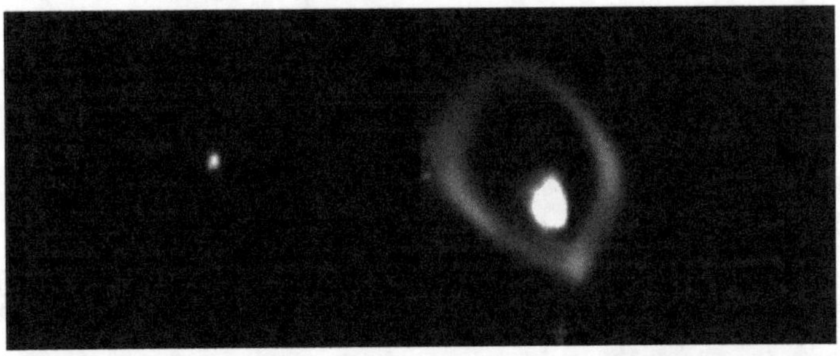

A MOT cloud in two different density regimes:If the density of the MOT is high enough, the MOT cloud goes from having a Gaussian density distribution (left), to something more exotic (right). In the right hand image, the density is so high that atoms have been blown out of the central trapping region by radiation pressure, to then form a toroidal racetrack mode around it.

The minimum temperature and maximum density of a cloud in a magneto-optical trap is limited by the spontaneously emitted photon in cooling each cycle. While the asymmetry in atom excitation gives cooling and trapping forces, the emission of the spontaneously emitted photon is in a random direction, and therefore contributes to a heating of the atom. Of the two $\hbar k$ *kicks the atom receives in each cooling cycle, the first cools, and the second heats: a simple description of laser cooling which enables us to calculate a point at which these two effects reach equilibrium, and therefore define a lower temperature limit, known as the* Doppler cooling limit.

The density is also limited by the spontaneously emitted photon. As the density of the cloud increases, the chance that the spontaneously emitted photon will leave the cloud without interacting with any further atoms tends to zero. The absorption, by a neighboring atom, of a spontaneously emitted photon gives a $2\hbar k$ *momentum kick between the emitting and absorbing atom which can be seen as a repulsive force, similar to coulomb repulsion, which limits the maximum density of the cloud.*

Application

As a result of low densities and speeds of atoms achieved by optical cooling, the mean free path in a ball of MOT cooled atoms is very long, and atoms may be treated as ballistic. This is useful for quantum information experiments where it is necessary to have long coherence times (the time an atom spends in a defined quantum state). Because of the continuous cycle of absorption and spontaneous emission, which causes decoherence, any quantum manipulation experiments must be performed with the MOT beams turned off. In this case, it is common to stop the expansion of the gases during quantum information experiments by loading the cooled atoms into a dipole trap.

A magneto-optical trap is usually the first step to achieving Bose–Einstein condensation. Atoms are cooled in a MOT down to a few times the recoil limit, and then evaporatively cooled which lowers the temperature and increases the density to the required phase space density.

A MOT of 133Cs was used to make some of the best measurements of CP violation.

References

- "Future Trends in Fiber Optics Communication" (PDF). WCE, London UK. July 2, 2014. ISBN 978-988-19252-7-5
- Roberto Scopigno; Susanna Bracci; Falletti, Franca; Mauro Matteini (2004). Exploring David. Diagnostic Tests and State of Conservation. Gruppo Editoriale Giunti. ISBN 88-09-03325-6
- Paul A. Strassmann (2008). The Computers Nobody Wanted: My Years with Xerox. Strassmann, Inc. pp. 126–. ISBN 978-1-4276-3270-8
- "Green lasers halt Coast Guard air searches "Coast Guard Compass". Coastguard.dodlive.mil. 2011-02-28. Retrieved 17, August 2020
- Michael Shawn Malone (2007). Bill & Dave: How Hewlett and Packard Built the World's Greatest Company. Penguin. pp. 327–. ISBN 978-1-59184-152-4
- Roy A. Allan (1 October 2001). A History of the Personal Computer: The People and the Technology. Allan Publishing. pp. 13–. ISBN 978-0-9689108-3-2
- Mary Kay Carson (2007). Alexander Graham Bell: Giving Voice To The World. Sterling Biographies. New York: Sterling Publishing. pp. 76–78. ISBN 978-1-4027-3230-0
- William E. Kasdorf (January 2003). The Columbia Guide to Digital Publishing. Columbia University Press. pp. 383–. ISBN 978-0-231-12499-7
- Edwin D. Reilly (2003). Milestones in Computer Science and Information Technology. Greenwood Press. ISBN 1-57356-521-0
- Taylor, Jim H.; Johnson, Mark R.; Crawford, Charles G. (2006). DVD Demystified. McGraw-Hill Professional. pp. 7–8. ISBN 0-07-142396-6
- S. Nagabhushana (2010). Lasers and Optical Instrumentation. I. K. International Pvt Ltd. pp. 269–. ISBN 978-93-80578-23-1

- "UK: Plane Bombs Explosions Were Possible Over U.S". Fox News. Archived from the original on March 29, 2012. Retrieved 17, May 2020
- He C, Morawska L, Taplin L (2012). "Particle emission characteristics of office printers. ;Environ Sci Technol. 2007] - PubMed - NCBI". ncbi.nlm.nih.gov. Retrieved 05, August 2012
- Murphy, Liam. "Case Study: Old Mine Workings". Subsurface Laser Scanning Case Studies. Liam Murphy. Retrieved 11, January 2020
- Ross, Selina (5 November 2015). "Laser pointers not toys, optometrists warn, after Tasmanian teenager damages eyes". ABC News. Australian Broadcasting Corporation. Retrieved 05, January 2020
- Metzger, Robert M. (2012). The Physical Chemist's Toolbox. John Wiley & Sons. p. 207. ISBN 978-0-470-88925-1. Retrieved 15, June 2020
- Gladwell, Malcolm (May 16, 2011). "Creation Myth - Xerox PARC, Apple, and the truth about innovation". The New Yorker. Retrieved 28, June 2020

Permissions

All chapters in this book are published with permission under the Creative Commons Attribution Share Alike License or equivalent. Every chapter published in this book has been scrutinized by our experts. Their significance has been extensively debated. The topics covered herein carry significant information for a comprehensive understanding. They may even be implemented as practical applications or may be referred to as a beginning point for further studies.

We would like to thank the editorial team for lending their expertise to make the book truly unique. They have played a crucial role in the development of this book. Without their invaluable contributions this book wouldn't have been possible. They have made vital efforts to compile up to date information on the varied aspects of this subject to make this book a valuable addition to the collection of many professionals and students.

This book was conceptualized with the vision of imparting up-to-date and integrated information in this field. To ensure the same, a matchless editorial board was set up. Every individual on the board went through rigorous rounds of assessment to prove their worth. After which they invested a large part of their time researching and compiling the most relevant data for our readers.

The editorial board has been involved in producing this book since its inception. They have spent rigorous hours researching and exploring the diverse topics which have resulted in the successful publishing of this book. They have passed on their knowledge of decades through this book. To expedite this challenging task, the publisher supported the team at every step. A small team of assistant editors was also appointed to further simplify the editing procedure and attain best results for the readers.

Apart from the editorial board, the designing team has also invested a significant amount of their time in understanding the subject and creating the most relevant covers. They scrutinized every image to scout for the most suitable representation of the subject and create an appropriate cover for the book.

The publishing team has been an ardent support to the editorial, designing and production team. Their endless efforts to recruit the best for this project, has resulted in the accomplishment of this book. They are a veteran in the field of academics and their pool of knowledge is as vast as their experience in printing. Their expertise and guidance has proved useful at every step. Their uncompromising quality standards have made this book an exceptional effort. Their encouragement from time to time has been an inspiration for everyone.

The publisher and the editorial board hope that this book will prove to be a valuable piece of knowledge for students, practitioners and scholars across the globe.

Index

www.ingramcontent.com/pod-product-compliance
Lightning Source LLC
LaVergne TN
LVHW080614210926
848485LV00003B/157